HUMAN-ROBOT INTERACTION IN
SOCIAL ROBOTICS

HUMAN-ROBOT INTERACTION IN
SOCIAL ROBOTICS

TAKAYUKI KANDA AND HIROSHI ISHIGURO

CRC Press
Taylor & Francis Group
Boca Raton London New York

CRC Press is an imprint of the
Taylor & Francis Group, an **informa** business

CRC Press
Taylor & Francis Group
6000 Broken Sound Parkway NW, Suite 300
Boca Raton, FL 33487-2742

First issued in paperback 2017

© 2013 by Taylor & Francis Group, LLC
CRC Press is an imprint of Taylor & Francis Group, an Informa business

No claim to original U.S. Government works

Version Date: 20120822

ISBN 13: 978-1-4665-0697-8 (hbk)
ISBN 13: 978-1-138-07169-8 (pbk)

Visit the Taylor & Francis Web site at
http://www.taylorandfrancis.com

and the CRC Press Web site at
http://www.crcpress.com

Contents

About the Authors

Takayuki Kanda received his bachelor's and master's in engineering, and Ph.D. degrees in computer science from Kyoto University, Kyoto, Japan, in 1998, 2000, and 2003, respectively. From 2000 to 2003, he was an intern researcher at ATR Media Information Science Laboratories, and he is currently a senior researcher at ATR Intelligent Robotics and Communication Laboratories, Kyoto, Japan. His current research interests include intelligent robotics, human–robot interaction, and vision-based mobile robots. Dr. Kanda was named to serve as a steering committee co-chair of ACM/IEEE international conference of human–robot interaction from 2010 to 2013.

Hiroshi Ishiguro received his doctoral degree in engineering from Osaka University, Japan, in 1991. In 1991, he started working as a research assistant in the Department of Electrical Engineering and Computer Science, Yamanashi University, Japan. Then, he moved to the Department of Systems Engineering, Osaka University, Japan, as a research assistant in 1992. In 1994, he became an associate professor in the Department of Information Science, Kyoto University, Japan, and started research on distributed vision using omnidirectional cameras. From 1998 to 1999, he worked in the Department of Electrical and Computer Engineering, University of California, San Diego, as a visiting scholar. He has been Visiting Researcher at ATR Media Information Science Laboratories since 1999, where he has developed the interactive humanoid robot, Robovie. In 2000, he moved to the Department of Computer and Communication Sciences, Wakayama University, Japan, as an associate professor and became a professor in 2001.

He is now a professor in the Department of Adaptive Machine Systems, Osaka University, and a group leader at ATR Intelligent Robotics and Communication Laboratories.

Preface

We thank the support provided by the Ministry of Internal Affairs and Communications of Japan. Their funding enabled us to establish the concept of "network robots," which worked as the technical base for the studies presented in the book.

Further, this book is enabled by several years of collaboration with many researchers. Particularly, Norihiro Hagita is one of the leading persons for the concept of "network robots," and without his conceptual leadership this book was not possible. We also wish to thank to the following colleagues: Takahiro Miyasita, Michita Imai, Tatsuya Nomura, Masahiro Shiomi, Dylan F. Glas, Satoru Satake, and Satoshi Koizumi.

One of the unique characteristics of the works presented in the book is the field studies. These studies were possible thanks to cooperation of people in the following facilities: Osaka Science Museum, Kinki Nippon Railway Co., Ltd., Takanohara AEON, and Universal City Walk, Osaka. We wish to thank all of them. We also wish to thank visitors of these facilities as well as participants of our laboratory studies. Their participation and feedback are essential to the research in human–robot interaction.

1

Introduction to Network Robot Approach for Human–Robot Interaction

1.1 From Navigation and Manipulation to Human–Robot Introduction

There are two main streams in the early robotics. One is navigation and the other is manipulation. Navigation is the main function of an autonomous mobile robot. It observes the environment with cameras and laser scanners and builds the environmental model. With the acquired environmental model, it makes plans to move from the starting point to the destination. This is the navigation, and it was the main issue not only for robotics but also for artificial intelligence. On the other hand, manipulation is the main function of a robot arm. Many robot arms are working for assembling products in industries. They need to handle various objects in complicated environments. Although the space for the manipulation is limited, the structure of the robot arm is complex like a human. Therefore, manipulation required sophisticated planning algorithms.

The applications for navigation and manipulation were in industry. In other words, robotics was for industrial robots. And it completely changed the production lines. Recently, many robots are working instead of human workers. That is, we have solved main research issues for the industrial robots and developed the practical systems. Of course, we have not completely solved the research issues. It is quite important and necessary to deeply study the issue and improve the developed algorithms. However, on the other hand, robotics needs a new research issue.

That is "Interaction." Industrial robotics developed key components for building more human-like robots, such as sensors and motors. From 1990 to 2000, Japanese companies developed various animal-like and human-like robots. Sony developed AIBO (Figure 1.1), which is a dog-like robot and QRIO (Figure 1.2), which is a small human-like robot. Mitsubishi Heavy Industries, LTD developed Wakamaru (Figure 1.3). Honda developed a child-like robot called ASIMO (Figure 1.4). Unfortunately, Sony and Mitsubishi Heavy Industries, LTD have stopped the projects but Honda is still continuing.

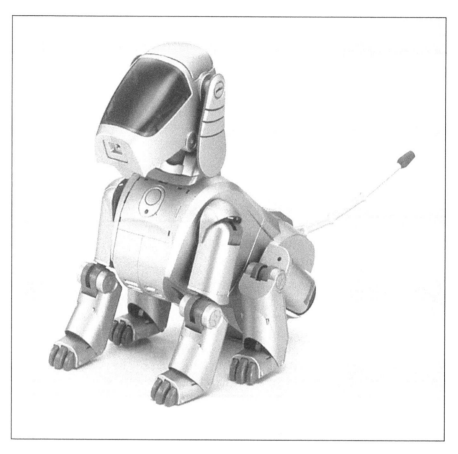

FIGURE 1.1
AIBO developed by Sony. (Source Sony Corporation.)

The purpose of these companies was to develop interactive robots with people. They have achieved a great success for the industrial robots. However, the development has been almost over, and they needed to develop new types of robots. Their focus was everyday life. They have tried to extend the application area of the robots from industrial environments to daily environments.

The most serious difference between industrial robots and robots works in the daily environments is "interaction." In the daily environments, robots encounter with humans, and they have to interact with them before doing the task. Rather, the interaction will be the main task of the robots. These robots developed by the companies had various functions with the animal-like or human-like appearance to interact with people.

For navigation, the target robot was vehicle-type robot or mobile platform. For manipulation, the target robot was manipulator or arm robot. For interaction, the target robot is humanoid that has a human-like shape.

FIGURE 1.2
QRIO developed by Sony. (Source Sony Corporation.)

FIGURE 1.3
Wakamaru developed by Mitsubishi Heavy Industries, LTD.

FIGURE 1.4
ASIMO developed by Honda.

1.2 Interactive Robots

Figure 1.5 shows our future society with interactive robots. The robots developed by the companies will become part of the futuristic scene shown in Figure 1.5. The strongest reason is in the human innate ability to recognize humans and prefer human interaction. The human brain does not react emotionally to artificial objects, such as computers and mobile phones. However, it has many associations with the human face and can react positively to resemblances to the human likeness. Therefore, the most natural communications media for humans are humans. That is, humanoids and androids that have a very human-like appearance will be ideal media for humans. This is the reason why some companies have developed humanoid models, and why many people are interested in humanoids.

However, it is not so easy to realize the robot world shown in Figure 1.5. We have to solve the following three issues:

a. Sensor network for tracking robots and people
b. Development of humanoids that can work in the daily environment.
c. Development of functions for interactions with people.

The robot cannot completely observe an environment with onboard sensors. The robot can have cameras and laser scanners and observe the local

FIGURE 1.5
Interactive robots working in our future society.

environment around itself. However, it is quite difficult to monitor the whole environment in real time with the sensors. On the other hand, the human functions in a more sophisticated way to monitor the environment with the limited sensors. Humans can guess and infer the events that happen in the environment based on experience. Therefore, even if the sensors have limitations, humans can widely monitor the environment.

The robot cannot take the same strategy as humans since it does not have the sophisticated method to infer events that happen in the wider environment with the onboard sensors. The current technology of artificial intelligence is not enough to simulate the human brain function.

The alternative way to monitor the wide environment is to use many sensors distributed in the environment. This is the idea of ubiquitous computing. We are already using various ubiquitous sensors in our current society. There are many surveillance cameras in banks, railway stations, and hospitals. Some of the cities are placing many cameras along streets for security purposes.

By extending this sensor network, we can develop more sophisticated sensor networks for monitoring both human and robot activities. Section 5 introduces the sensor network developed by us. It covers a wide area and stably monitors both humans and robots in real time.

The second issue is the development of humanoids. The most popular humanoid is ASIMO developed by Honda. The perfection level is very high, and it functions almost practical level. However, the biped mechanism is not perfectly safe. ASIMO cannot physically interact with young children. A more reliable and safe mechanism for moving is the mobile platform with two driving wheels used for Wakamaru.

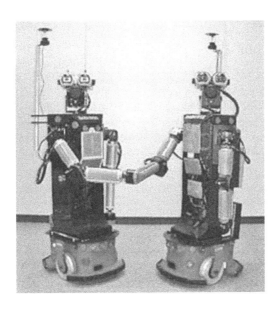

FIGURE 1.6
Robovie developed by ATR.

Wakamaru has been developed based on Robovie, an innovation from Intelligent Robotics and Communication Laboratories in Advanced Tele-communications Research Institute International (ATR). The design concept of Robovie was to have safe and stable hardware for interactive communication using gestures (Figure 1.6). It has two driving wheels and two arms and a head. Each arm has four degrees of freedom with minimum torque motors for gestures. Therefore, children can easily and physically stop the arm movements to avoid any dangerous situations.

In addition to the minimum mechanical hardware, Robovie has various sensors. It has two high-speed pan-tilt cameras on the head and an omnidi-rectional camera on the top of the pole attached in the back. The high-speed pan-title cameras are good for watching people standing in front of the robot, but they are not sufficient for observing the surrounding of the robot. For such a purpose, the omnidirectional camera works very well. Robovie is always monitoring its surroundings and quickly pays attention with the high-speed pan-title cameras when it finds something of significance.

Another important sensor is the tactile sensors that cover the whole body. Robovie also has several ultra-sonic sensors to monitor the surroundings. However, they are not enough. During interactions with people, they touch the body. In order to detect touching, we have developed tactile sensors by using conductive sheets.

The third issue is to develop functions for interacting with people. This is the most important issue for human–robot interaction. This book introduces many interactive behaviors implemented onto Robovie and the design policies.

1.3 Network Robots

The robots supported by the sensor networks are so-called *network robots*. The basic idea of the network robots is to enhance the robot abilities by integrating with ubiquitous network technology. The development of the network robots is supported by the Japanese government as a national project from 2004. The core members are ATR, NTT, Toshiba, Mitsubishi Heavy Industries, LTD, and Panasonic. The main purpose of the project is to extend the ability to recognize the real world and the ability to converse with a robot by connecting it with the sensor networks and other robots.

The most important progress of the project is the conversation ability of the robot. In addition to the laboratory experiments, we are performing various field tests of the developed robot in real environments. Through the field tests, we are verifying the conversation ability of the robots and finding new research issues.

This field tests are the most important methodology in the development of network robots. Without the field test, we cannot find real issues that we have to solve for realizing practical interactive robots. The fields for trials are elementary schools, stations, science museums, and shopping malls. In the field, we have considered how to integrate robots and sensor networks and what kinds of services are possible for the robots.

This book starts with introduction of the field tests and then discusses the necessary technologies. We authors hope this book provides rich information to readers for developing practical interactive robots.

2

Field Tests—Observing People's Reaction

2.1 Introduction

The fundamental issues of human–robot interaction are in the real world. The human–robot interaction is not for developing robots in a well-controlled situation like psychological studies. The purpose is to realize real services by the robot through the interactions. Therefore, it is quite important to have field tests and to design the interactive behaviors of the robot in the field test.

We researchers have mainly focused on the development of the general functions and behavior of the robot in laboratories. These would include moving to a destination and greeting. However, laboratories are not real environments, especially for service robots. The robot should have situated behaviors to the environment, and they should follow contexts. There would be the same as a human. Even if we develop many general functions or behaviors in the robot, the robot cannot properly behave without contextual information and recognition of a situation.

In order to solve this problem, we have started field tests of the interactive robot in elementary schools. We believe this is the first trial in the world. In the first field test, we have prepared a tag system for monitoring the relationship between robots and students. All students put tags on the name badges and the robot carried tag readers.

In this field test, we could understand how students establish relationships with the robots and other students. However, we needed to take more detailed information on the students' behaviors. And we needed to give the robot more active behaviors for moving around the environment to perform various tasks.

For a second field test, we have selected a science museum. In the science museum, we have installed many tags in the environment in addition to visitors' name badges. The tags installed in the environment were for localizing and navigating the robot. Further, we have installed many cameras in the environment and developed a tele-operated system for the robot.

In a complicated environment, it is quite difficult to design robot behaviors a priori. We cannot guess everything that might happen in the real world. Therefore, we have developed the tele-operated system. An operator remotely controlled the robot, and we recorded the robot sensory data and sensory data from the tags and cameras installed in the environment. We have tried to develop robot autonomous functions based on the recorded sensory data.

However, the merit of the tele-operated system was not only for developing robot autonomous behaviors in a complicated environment. The sensory data from both of the robot and environment taught us many things about how people react to robot behavior. It provided us rich information about human interactive behaviors, but there is a limitation to the psychological studies done in laboratories. They cannot provide enough variety of human interactive behaviors since there is no rich context in the laboratory environment. It is a rather assumed situation and not real. However, field tests with the tele-operated robot can induce real and rich interactive behavior among people. Thus, the tele-operated systems and the sensor network in the field test enable us to deeply study the psychological and cognitive aspect of people's behaviors.

This section introduces several field tests as a start to this book. Through the field test, we hope the readers will come to understand the real issues for interactive robots.

2.2 Interactive Humanoid Robots for a Science Museum

Masahiro Shiomi, Takayuki Kanda, Hiroshi Ishiguro, and Norihiro Hagita

ABSTRACT

This section reports on a field trial with interactive humanoid robots at a science museum where visitors are encouraged to study and develop an interest in science. In the trial, each participating visitor wore an RFID tag while looking around the museum's exhibits. Information obtained from the RFID tags was used to direct the robots' interaction with the visitors. The robots autonomously interacted with visitors via gestures and utterances resembling the free play of children [1]. In addition, they guided visitors around several exhibits, and explained the exhibits based on sensor information. The robots were given high evaluations by visitors during the 2-month trial. In addition, we conducted an experiment during the field trial to compare in detail effects of exhibit-guiding and free-play interaction under three operating conditions. The results revealed that the combination of free-play interaction and exhibit-guiding positively affected visitors' experiences at the science museum.

Index Terms: Robotics, field trial, interactive robot, intelligent systems, science museum

2.2.1 Introduction*

Our objective is to develop an intelligent communication robot that operates in a large-scale daily environment such as a museum to support people through interactions with body movements and speech. We have selected a humanoid robot to achieve our objective because its physical structure enables it to interact with people using human-like body movements such as shaking hands, greeting, and pointing. Such interactions are more likely to be understood by both adults and children than interaction with an electronic interface such as a touch panel or buttons. In addition, possessing a human-like body is useful for naturally holding the attention of people [2]. We expect human-like interaction to be important for improving the perceived friendliness of the robot toward people.

To behave intelligently during an interaction, the robot requires many types of information about its environment and the people with whom it is interacting. For example, it is reasonable to suppose that people interacting with the robot would expect some human-like interaction such as greeting by name. It is difficult, however, for a robot to obtain such information in a daily environment; particularly, in a large-scale daily environment such as a museum, simple functions such as person identification are very difficult because the robot's sensing ability is likely to be affected by the presence and changing movement of a large number of people, as well as unfavorable illumination and background conditions in the environment.

To achieve our objective, we integrate autonomous robotic systems and ubiquitous sensors that support robots. This approach involves applying ubiquitous sensors to monitor the environment and to acquire rich sensory information and process it, enabling an autonomous robot to freely interact with people by utilizing the sensory data received from the ubiquitous sensors. For example, a robot would choose behaviors based on certain personal information about people such as name and movement history, which is achieved by distributed RFID tags and readers. This approach enables the robot to provide more pertinent information during interaction, such as recommendations based on each visitor's movement history. Moreover, the cameras supply accurate coordinates to the robot that are used for localizing and navigating the robot. We explore the potential of communication robots with this approach.

---

* This chapter is a modified version of a previously published paper by Masahiro Shiomi, Takayuki Kanda, Hiroshi Ishiguro, and Norihiro Hagita, *Interactive Humanoid Robots for a Science Museum*, ACM/IEEE 1st Annual Conference on Human-Robot Interaction (HRI2006), pp. 305–312, 2006, edited to be comprehensive and fit with the context of this book.

TABLE 2.1

Various Field Experiments with Interactive Robots

		Interaction		Function	
Purpose	Location	Human-Like	Using Personal Information	Person Identification	Navigation
Mental care [3]	Hospital	—	—	—	—
Language education [4]	School	✓	✓	✓	—
Assistant [5]	Nursing homes	—	—	—	—
Guidance and navigation [6]	Museum	—	—	—	✓
Guidance and navigation [7]	Expo	—	—	✓	✓
Interaction and guidance [This chapter]	**Museum**	✓	✓	✓	✓

2.2.2 Related Works

Table 2.1 shows a comparison between our work and previous works based on the concept of the communication robot from four points of view: purpose, experimental environment, details of interactions, and functions.

Aibo and Paro [3] had animal-like appearance—respectively, a dog and a seal. These robots provide entertainment or mental care to people through a human–pet style of interaction. Both studies indicated the effectiveness of interaction between people and pet-like robots.

Robovie [4] was used to assist with language education in elementary school. This research detailed the importance of using personal information in an interaction. However, Robovie only interacted with a limited group of people; thus, it is not clear how a robot should operate in large-scale environments where a wide variety of people visit.

Nursebot [5], RHINO [6], and RoboX [7] are traditional mobile robots, the developers of which designed robust navigation functions for daily environments. In particular, RoboX and RHINO guided thousands of people in large-scale environments. Although these works represent the effectiveness of robust navigation functions in interactions between people and robots, their interaction functions are quite different from human-like interactions.

To create an intelligent communication robot that can operate in a large-scale environment, we consider it important to investigate the effectiveness of human-like interaction as well as developing robust functions. Therefore, we designed our robots to interact with people using human-like bodies and personal information obtained via ubiquitous sensors.

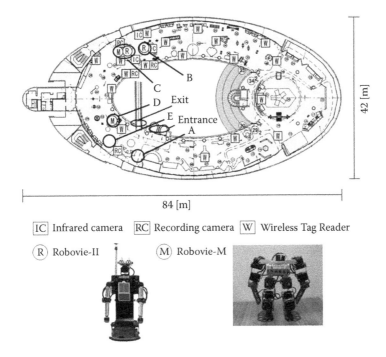

IC Infrared camera   RC Recording camera   W Wireless Tag Reader

R Robovie-II   M Robovie-M

FIGURE 2.1
Map of the fourth floor of the Osaka Science Museum

2.2.3 The Osaka Science Museum

2.2.3.1 General Settings

Seventy-five exhibits were positioned on the fourth floor of the Osaka Science Museum. Visitors could freely explore the exhibits. Figure 2.1 shows a map of the fourth floor of the museum. Generally, visitors walk in the counterclockwise direction from the entrance to the exit. The width and length of the Osaka Science Museum are 84 [m] and 42 [m], respectively.

2.2.3.2 Our Experimental Settings

We installed the humanoid robots, RFID tag readers, infrared cameras, and video cameras in the Osaka Science Museum for experiments. Visitors could freely interact with our robots similar to the other the exhibits. Typically, in our experiment, visitors progress through the following steps:

- If a visitor decides to register as part of our project, personal data such as name, birthday, and age (under 20 or not) is gathered at the reception desk (Figure 2.1, point A). The visitor receives a tag at the reception desk. The system binds those data to the ID of the tag and automatically produces a synthetic voice for the visitor's name.

- The visitors can freely experience the exhibits in the Osaka Science Museum as well as interact with our robots. Four robots are placed at positions B, C, and D on the fourth floor, as shown in Figure 2.1.

After finishing, visitors return their tags at the exit point (Figure 2.1, point E).

2.2.4 System Configuration

Our intelligent and interactive system at the Osaka Science Museum consists of four humanoid robots and ubiquitous sensors.

Figure 2.2 represents an overview of the system. The upper part represents the ubiquitous sensors. The lower part represents robots; the details are covered in section 2.2.4.3.1. The robots reacted to information from their sensors, as well as information from the ubiquitous sensors. The ubiquitous sensors recorded the movements and positions of visitors on the fourth floor of the Osaka Science Museum via their RFID tags. These sensors were also used to identify visitors and estimate the correct coordinates of the robots. The visitors' information was used for interaction between robots and visitors. The interaction data between robots and visitors were recorded in a database using recording cameras. Generally, the robots behaved as follows:

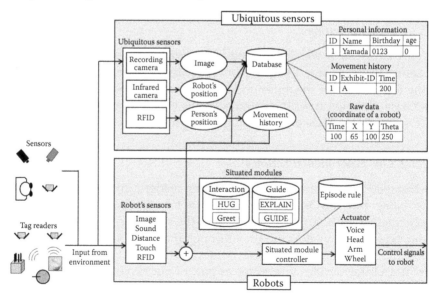

FIGURE 2.2
An overview of our system.

One robot served as a guide to the exhibits.

Two stationary robots explained the exhibits.

As visitors prepared to leave, one robot greeted them by name, asked them to return their RFID tags, and said goodbye.

2.2.4.1 Database

All ubiquitous sensor data and robots' sensor data were recorded in the database with a time stamp; personal information and movement histories were also recorded. These data were used for deciding robots' behavior, interacting with visitors, and analyzing the results of our experiments. Details of recorded data are as follows:

From reception: IDs of the visitors' RFID tags, registered visitors' personal information, and times of registration/return tags

From RFID tag readers: Times when visitors approached the particular exhibits and estimated visitors' positions

From all ubiquitous sensors and robots: Each sensor's output.

2.2.4.2 Embedded Ubiquitous Sensors in an Environment

On the fourth floor of the Osaka Science Museum, we installed 20 RFID tag readers (Spider-IIIA, RF-CODE, which included the two fitted to the Robovies), three infrared sensors, and four video cameras. All sensor data were sent to the central database through an Ethernet network.

In the following subsections, we describe each type of sensor used.

RFID Tag Readers

We used an active type of RFID tag, a technology that enables easy identification of individuals. Detection is unaffected by the occurrence of occlusions, the detection area is wide, and the distance between the tag reader and an RFID tag can be roughly estimated. Such benefits make it suitable for large environments. Furthermore, RFID tag readers are used for recording the time when each tagged visitor approached particular exhibits and for estimating the visitors' positions. We placed the readers around particular exhibits to detect whether visitors approached them. Figure 2.1 shows the positions of the tag readers.

Infrared Cameras

We placed an infrared LED on top of Robovie and attached infrared cameras to the ceiling in order to correct the robot's position (to accurately navigate a robot in a crowded museum). The system produces binary images from the infrared cameras and detects bright areas. It then calculates absolute coordinates with reference to the weighted center of the detection area and sends them to the database. Infrared camera positions are shown in Figure 2.1. The distance between the floor and the ceiling is about 4 m. The width and height of images from an infrared camera is 320 and 240 pixels, respectively. One pixel represents about 1 cm² of area.

Video Cameras

The video camera positions are also shown in Figure 2.1. The output images of each video camera are recorded onto a PC and used to analyze the data generated during the experiment.

2.2.4.3 Humanoid Robots

We used two types of humanoid robot: Robovie and Robovie-M. This section provides details of the robots.

2.2.4.3.1 Robovie

Hardware "Robovie" is an interactive humanoid robot characterized by its human-like physical expressions and its various sensors. The reason we used humanoid robots is because a human-like body is useful for naturally holding the attention of humans [2]. Its height is 120 cm, and its diameter is 40 cm. The robot has two 4 × 2 DOFs in its arms, three DOFs in its head, and a mobile platform. It can synthesize and produce a voice via a speaker. We also attached an RFID tag reader to Robovie that enables it to identify individuals around it [4]. In this system, we use Robovies as sensors because they each contain an RFID tag reader. In effect, they became not only interactive robots but also part of the sensor system.

Situated Modules Robovie's software regime comprises *situated modules*, a *situated module controller*, and *episode rules* [8], which are used to perform consistent interactive behaviors. Interactive behaviors are designed with knowledge about the robot's embodiment, obtained from cognitive experiments and then implemented as *situated modules* with situation-dependent sensory data processing for understanding complex human behaviors. *Situated modules* realize certain interactive behaviors such as calling a visitor's name, shaking hands, greeting, explaining exhibits, and guiding visitors around exhibits. In our system, the *situated modules* were designed together with the robot's sensors and the ubiquitous sensors.

The relationships among behaviors are implemented as rules governing the execution order (named *episode rules*) to maintain a consistent context for communication. *The situated module controller* selects a situated module based on *episode rules* and sensory data. Figure 2.3 shows one example of *episode rules* and scenes of an interaction between a robot and a visitor. The episode rules were designed to guide robot behavior, depending on recorded data in the database and outputs of the robots' sensors. In this case, the

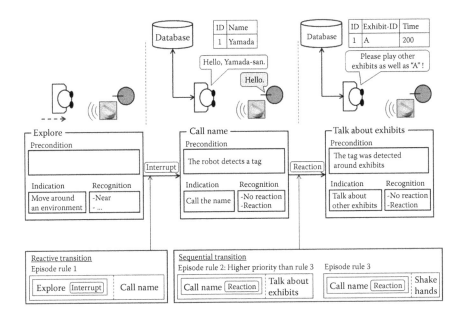

FIGURE 2.3
An illustration of episode rules and scenes of an interaction

robot explores the environment by "Explore," then detects a visitor's tag. This causes a reactive transition ruled by episode rule 1. The robot calls the visitor's name by executing the situated module "Call name" (Figure 2.3, top center). After executing "Call name," it starts talking about exhibits by executing "Talk about exhibits" (Figure 2.3, top right). This sequential transition is caused by episode rule 2, which has a higher priority than episode rule 3.

The number of situated modules developed to date has reached 230: 110 are interactive behaviors such as hand-shaking, hugging, playing paper-rock-scissors, exercising, greeting, and engaging in a short conversation; 40 are idling behaviors such as the robot scratching its head and folding its arms; 20 are moving-around behaviors such as pretending to patrol an area; 50 are explaining and guiding behaviors for exhibits; and 20 are talking behaviors based on information from an RFID tag. The number of episode rules for describing relationships among these modules exceeds 800.

2.2.4.3.2 Robovie-M

"Robovie-M" is a humanoid robot characterized by its human-like physical expressions. We decided on a height of 29 cm for this robot. Robovie-M has 22 DOFs and can perform two-legged locomotion, bow its head, and do a handstand. We used a personal computer and a pair of speakers to enable it to speak, since it was not originally equipped for that. The behavior of each Robovie-M is decided by the behavior of a Robovie or output from the RFID tag readers.

2.2.5 Robot Behavior

2.2.5.1 Locomotive Robot

We used a Robovie for the locomotive robot, which moved around in parts of the environment, interacted with visitors, and guided them to exhibits. Each robot's behavior is automatically decided by episode rules that are based on data from its sensors and the database. Such behavior can be divided into four types, the details of which are as follows:

Interaction with Humans: Childlike Interaction

The robot can engage in such child-like behavior as handshaking, hugging, and playing the game of "rock, paper, and scissors." Moreover, it has such reactive behaviors as avoidance and gazing at a touched part of its body, as well as such patient behavior as solitary playing and moving back and forth. Figure 2.4a shows interaction scenes between Robovies and visitors.

Interaction with Humans: Using Information from RFID Tags

The robots can detect RFID tag signals around themselves by using their RFID tag reader, which allows them to obtain personal data on visitors using RFID tag IDs. Each robot can greet visitors by name or wish them a happy birthday, and so on. In addition, the system records the time that visitors spend on the fourth floor of the Osaka Science Museum. The robots can behave according to that time.

Guiding People to Exhibits: Human Guidance

The robot can guide people to four kinds of exhibits by randomly determining the target. Figure 2.4b,c shows an example of this behavior. When bringing visitors to the telescope, the robot says, "I am taking you to an exhibit, please follow me!" (b-1), and approaches the telescope (b-2, 3). It suggests that the person look through it and then talks about its inventor (b-4).

Guiding People to Exhibits: Using Information from RFID Tags

The RFID tags' data are also used for interaction. We used the amount of time that visitors spent near an exhibit to judge whether visitors tried it. For example, when an RFID-tagged visitor has stayed around the Magnetic Power exhibit longer than a predefined time, the system assumes that the visitor has already tried it. Thus, the robot says, "Yamada-san, thank you for trying Magnetic Power. What did you think of it?" If the system assumes that the visitor has not tried it, the robot will ask, "Yamada-san, you didn't try Magnetic Power. It's really fun, so why don't you give it a try?"

2.2.5.2 Robots That Talk with Each Other

Two stationary robots (Robovie and Robovie-M, Figure 2.4d-1,2) can casually talk about the exhibits as humans do with accurate timing because they

(a-1)

(a-2)

(a-3)

FIGURE 2.4

Scenes of interaction between visitors and our robots: (a) Scenes of interaction between visitors and Robovie; (b) Robovie guiding visitors to the telescope; (c) scenes when visitor had an interest in the telescope: (1) two robots talking, (2) two robots talking to visitors, (3) the visitor talking to the robot, (d) scenes of stationary robots interacting with visitors.

(b-1)

(b-2)

(b-3)

FIGURE 2.4 (continued)

(b-4)

(c-1)

(c-2)

FIGURE 2.4 (continued)

(d-1)

(d-2)

(d-3)

FIGURE 2.4 (continued)

are synchronized with each other using an Ethernet network. A behavior of Robovie was decided by episode rules based on data from its RFID tag reader and the database. Furthermore, Robovie controlled timing of Robovie-M's motion and speech. The topic itself is intelligently determined by data from RFID tags. By knowing a visitor's previous course of movement through the museum, the robots can try to interest the visitor in an exhibit he or she overlooked by starting a conversation about that exhibit.

2.2.5.3 A Robot Saying Goodbye

This robot is positioned near the exit and, after requesting data from their RFID tags, says goodbye to the departing visitors. It also reorients visitors on the tour who are lost by examining the visitor's movement history and time spent on the fourth floor of the Osaka Science Museum, which was recorded by the system. If visitors walk clockwise, they will immediately see this robot at the beginning and will be pointed in the right direction by the robot. Figure 2.4d-3 shows a scene with this robot.

2.2.6 Experiment

2.2.6.1 A 2-Month Exhibition

We performed experiments to investigate the impressions made by robots on visitors to the fourth floor of the Osaka Science Museum during a two-month period. The questionnaire used in the exhibition consisted of the following statements. Respondents indicated the degree to which each statement applied on a scale of 1 to 5:

(1) "Interesting": I am interested in the robots.
(2) "Friendly": I felt friendly toward the robots when I faced them.
(3) "Effective": I found the guidance provided by the robots was effective.
(4) "Anxiety about interaction": I felt anxious when the robots talked to me.
(5) "Anxiety about future robots": I feel anxious about the possible widespread application of robots to perform tasks in the near future such as those shown at the exhibition.

2.2.6.2 Results of the 2-Month Experiment

By the end of the two-month period, the number of visitors had reached 91,107, the number of subjects who wore RFID tags was 11,927, and the number of returned questionnaires was 2,891. The results of the two-month experiment indicate that most visitors were satisfied with their interaction with our robots. In this section, we describe three kinds of results: the results

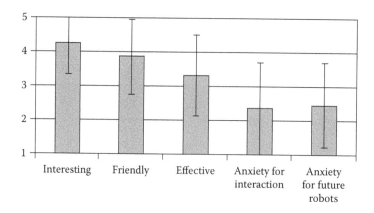

FIGURE 2.5
Results of returned questionnaires.

of each item on the questionnaire; freely described opinions; and observed interactions between visitors and our robots.

2.2.6.2.1 Results for Each Item in the Questionnaire

Figure 2.5 shows the results and averages from the questionnaires, providing three findings. The first is that most visitors had an interest in and good impressions of the robots because the score for "Interesting" was above four and that for "Friendly" reached four. The second finding relates to the effectiveness of guidance. The score for "Effective" was nearly three, indicating that the robots' guidance did not leave a particularly strong impression on visitors. The third is that visitors did not feel anxious about interaction with these robots or with future robots. These findings are supported by the scores for "Anxiety about interaction" and "Anxiety about future robots" (both scores are less than the middle value). Altogether, the results from the questionnaires indicate that most visitors had interest in and good impressions of our robots, and did not feel anxious about them.

2.2.6.2.2 Freely Described Opinions

The results revealed that visitors held favorable impressions toward the presence of the robots. Moreover, visitors described their favorite robot behavior, such as hugging, the calling out of names, and so on. Such behaviors are basic elements of human society. Most opinions were along the lines of:

We had a really good time.

I had fun because the robots called me by name.

We felt close to the robots.

The freely described opinions of visitors were analyzed, and revealed that visitors' opinions of the robots differed according to age [9]. For example, younger respondents did not necessarily like the robots more than older respondents.

2.2.6.2.3 Observed Interaction between Visitors and Our Robots

We observed some interesting scenes between visitors and our robots during the experiment at the exhibition. We introduce the scenes as evidence that our robots interacted well with visitors.

2.2.6.2.3.1 Locomotive Robot

- Often there were many adults and children crowded around the robot. In crowded situations (mainly), a few children simultaneously interacted with the robot.
- Similar to Robovie's free-play interaction in a laboratory [1], children shook hands, played the paper-rock-scissors game, hugged, and so forth. Sometimes they imitated the robot's body movements, such as the robot's exercising.
- When the robot started to move to a different place (in front of an exhibit), some children followed the robot to the destination.
- After the robot explained a telescope exhibit, one child went to use the telescope (Figure 2.4c). When she came back to the robot, another child used the telescope.
- Its name-calling behavior attracted many visitors. They tried to show the RFID tags embedded in the nameplates to the robot. Often, when one visitor did this, several other visitors began showing the robots their nameplates, too, as if they were competing to have their names called.
- A visitor reported that when the robot moved to him, he thought that it was aware of him, which pleased him.

We demonstrated that robots could provide visitors with the opportunity to play with and study science through exhibits they might have otherwise missed. In particular, it reminds us of the importance of making a robot move around, a capability that attracts people to interact with it. Moreover, as shown in the scene where children followed the locomotive robot, it drew their attention to the exhibit, although the exhibit (a telescope) was relatively unexciting. (The museum features many attractive exhibits for visitors to move and operate to gain an understanding of science, such as a pulley and a lever.)

2.2.6.2.3.2 Robots That Talk to Each Other

- There were two types of typical visitors' behavior. One was just to listen to the robots' talk. For example, after listening to them, the

visitors talked about the exhibit that was explained to them, and
sometimes visited the exhibit.

- The other is to expect to have their name called. In this case, the visitors paid rather less attention to the robots' talk, and instead showed their name to the robots, which is similar to the actions observed around the locomotive robot. Often, visitors would leave the front of the robot just after his/her name was called.

One implication is that displaying a conversation between robots can attract
people and convey information to them, even though the interactivity is very
low. Such examples are also shown in other work [10].

2.2.6.2.3.3 Robot Bidding Farewell

- There were two types of typical visitors' behavior. One was just to watch the robot's behavior.

- The other was, again, expectation to have their name called. In this case, the visitors often showed their name to the robots.

The cost of Robovie-M is far cheaper than that of Robovie-II. Although its
functionality is very limited, such as its small size, no embedded speech
functions (we placed a speaker nearby), and no sensors (an RFID reader was
also placed nearby), it entertained many visitors. Particularly, the effectiveness of the name-calling behavior was again demonstrated, as seen in the
children's behavior of returning their RFID tags.

2.2.6.3 Experiments on the Behavior of Robots

We expected that free-play interaction would be affected by the effectiveness of robots' services. We performed experiments to examine the behavior
of robots under three operating conditions during one week. We randomly
switched conditions between the morning and the afternoon. The subjects
were the visitors who had RFID tags and played with the robots. After their
interaction ended, we asked them to fill out a questionnaire in which they
rated three items on a scale of 1 to 7, where 7 is the most positive.

The items were "Presence of the robots" (What did you think about the
presence of robots in the science museum?), "Usefulness as a guide" (What
was the degree of the robots' usefulness for easily looking around the exhibits?), and "Experience with science and technology" (How much did the
robots increase your interest in science and technology?). The subjects were
also encouraged to provide other opinions about the robots as well. The
three operating conditions were the following:

Interaction

Robots behaved according to predefined functions. Each robot engaged in basic
interaction, as described in Section 2.4.7.1. No guide function was performed.

Guidance

The role of the robots was limited to guiding and giving explanations. Each robot only behaved as described in Section 2.4.7.3.

Interaction, Guidance, and Using RFID Tags

In this operating condition the robots not only combined the previous two operating conditions but also used data from the RFID tags. Each robot performed every kind of behavior introduced in Section 2.4.7.

It is difficult to compare the conditions of "using RFID" and "not using RFID." For example, in the "Guide" condition, using information on the RFID tag necessitates that the robot behave interactively, such as calling someone by name. Also, it is difficult to compare the effects of each behavior in crowded situations. Thus, we use this operating condition for comparing the importance of the information on the RFID tags among the above conditions.

2.2.6.4 Results of Robots' Behavior

About 100 questionnaires were returned for each operating condition. Figure 2.6 shows the results and their averages, which are mostly above 6. There was a significant difference for the following item, "Experience with science and technology," as to whether the robot was in the "Interaction, guidance, and using RFID" operating condition or in another condition ($p < .05$).

A comparison of the three conditions' results based on analysis of variance revealed no significant differences between the two items of "Presence of the robots" and "Usefulness as a guide."

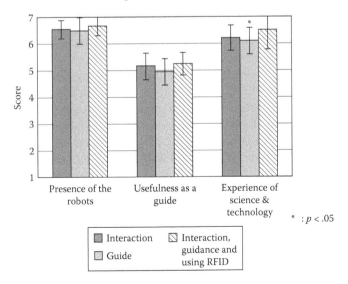

FIGURE 2.6
Results for the three operating conditions.

Concerning this last item, here are examples of some of the most remarkable feedback:

Children developed an interest in other exhibits after being led to them or having them explained by a robot.

Children were amused by the robot's reactions to being touched and became interested in new exhibited items when following it.

These opinions indicate that interest in science can be developed by possible interaction with robots. Other feedback opinions attest to the good impressions that robots made on subjects.

On the other hand, there were instances where robots could not interact well with visitors. For example, some children were afraid to interact with the robots, and some visitors did not care about the robots' actions. Moreover, visitors' opinions included some negative impressions such as "We couldn't talk to the robots because the robots' speech ability was not good." These indicate that the robot's interaction ability was not good enough for an open environment.

2.2.7 Discussion and Conclusion

2.2.7.1 Contributions

We believe that this trial demonstrated the high potential of using interactive humanoid robots in open environments, which is one of the most important contributions of this work. More than 90,000 people visited the exhibition, more than 10,000 interacted with the robots and wore RFID tags, and about 3,000 people returned questionnaires. The questionnaire results showed that most visitors evaluated the robots highly.

In addition, this trial revealed one appropriate design of interactive robots in a science museum. Human-like guiding with childlike free-play interaction attracted the interest of visitors more than simple guiding did. We believe that free-play interaction and human-guide-like guiding improved the visitors' interest in science and technology because the visitors played actively with the robots and with the exhibits via interactions with the robots.

In future work, we will try to improve interactive humanoid robots through findings from this field trial and do field experiments in other environments with the robots to investigate the possibility of communication robots.

2.2.7.2 A Perspective on Autonomy for a Communication Robot in Daily Life

Here, we discuss the essential technologies that made our approach possible, and describe how they were used and how we can further exploit them.

2.2.7.2.1 Humanoid Robot

The field trial revealed advantages of using humanoid robots when interacting with people in a daily environment. In the exhibition, we often observed that people stopped at the robots, saw the robots, and started to interact with the robots. Children, in particular, often did that. In other words, the presence of a humanoid robot attracted people to interact with it. This is one of the advantages of the humanoid robots.

Moreover, a human-like body enables the robots to interact with people in ways that are similar to the way humans communicate, such as shaking hands, greeting, and pointing. These behaviors made their interaction more natural. For instance, when a robot asked to shake hands and greeted visitors by calling their names, they also shook hands and greeted the robot. When it guided them to an exhibit and explained it with pointing gestures, children followed and started to learn about the exhibit.

Related to this, its tactile interaction capability was effective, which supplemented its lack of vision and auditory communication capability. For instance, when it asked to shake hands, it recognized the person's reaction with the tactile sensor attached to its hand. Similarly, visitors were allowed to touch robots' shoulders to start interacting with them; when the robot were touched, they turned to the visitors, greeted them, and asked to shake hands. We believe that these human-like behaviors of the robot and stable reactions with tactile sensors made people feel that the robot was more autonomous and realistic, which enabled them to engage deeply in the interactions.

On the other hand, the lack of speech recognition capability is one important problem. Although it spoke much about the exhibits, the robot could not listen to the questions and comments from the visitors. When a reliable technique of speech recognition in a noisy environment becomes available, the robots will be more useful and strong in this kind of application. Nevertheless, it is interesting that people already accepted and appreciated the robot even without the speech recognition capability.

2.2.7.2.2 RFID

We believe that person identification is one of the essential parts of human–robot communication. A useful human identification system needs to be robust. Mistakes in identification spoil communicative relationships between the robot and humans. For example, if a robot talks with a person and uses another person's name, it will negatively impact their relationship. We used RFID to realize robust person identification, which enabled us to identify multiple people at the same time. This would have been very difficult using only visual and auditory sensors in noisy real world.

As shown in the scenes of interaction with the robots, the person identification with RFID greatly promoted the human–robot interaction; the name-calling behavior, in particular, had a great impact. For instance, people were often crowded around the robots and showed their tags to the robot to have their name called by the robots. On the contrary, the information obtained from the RFID readers distributed to the whole environment made a relatively small contribution to the system. Robots talked to the visitors about their exhibit-visiting experience, such as "You did not see the telescope exhibit, did you? It is very interesting. Please try it," based on the information from the distributed RFID readers, but it seemed to be less impressive to the visitors. Perhaps robots are too novel for visitors; while they highly appreciated their experience of interacting with the robots, less attention was paid to the actual detailed services they offer.

2.2.7.2.3 Other Ubiquitous Sensors

The infrared camera supplied the exact position of the robot, which was very helpful in the crowded environment. We were able to fully exploit its presence owing to the robust navigation, such as moving around and guiding people to exhibits.

There are more ubiquitous sensors that will improve autonomous capabilities of communication robots as well as the infrared camera. We believe that ubiquitous sensors have excellent potential for providing data for human–robot interaction, and research on this will be included in our future work.

Acknowledgments

We wish to thank the staff at the Osaka Science Museum for their kind cooperation. This research was supported by the Ministry of Internal Affairs and Communications of Japan.

References

1. Ishiguro, H., Imai, M., Maeda, T., Kanda, T., and Nakatsu, R., Robovie: An interactive humanoid robot, *Int. J. Industrial Robot*, Vol. 28, No. 6, pp. 498–503, 2001.
2. Imai, M., Ono T., and Ishiguro, H., Physical relation and expression: Joint attention for human-robot interaction, *Proceedings of the 10th IEEE International Workshop on Robot and Human Communication (RO-MAN2001)*, pp. 512–517, 2001.
3. Shibata, T., An overview of human interactive robots for psychological enrichment, *The Proceedings of IEEE*, November 2004.
4. Kanda, T., Hirano, T., Eaton, D., and Ishiguro, H., Interactive robots as social partners and peer tutors for children: A field trial, *Journal of Human Computer Interaction*, Vol. 19, No. 1–2, pp. 61–84, 2004.

5. Pineau, J., Montemerlo, M., Pollack, M., Roy, N., and Thrun, S., Towards robotic assistants in nursing homes: Challenges and results. *Robotics and Autonomous Systems*, Vol. 42, Issues 3–4, 31, pp. 271–281.

6. Burgard, W., Cremers, A. B., Fox, D., Hähnel, D., Lakemeyer, G., Schulz, D., Steiner, W., and Thrun, S., The interactive museum tour-guide robot, *Proceedings of the National Conference on Artificial Intelligence (AAAI)*, 1998.

7. Siegwart, R. and et al. Robox at Expo.02: A large scale installation of personal robots, *Robotics and Autonomous Systems*, 42, 203–222, 2003.

8. Kanda T., Ishiguro H., Imai M., Ono T., and Mase K., A constructive approach for developing interactive humanoid robots, *IEEE/RSJ International Conference on Intelligent Robots and Systems (IROS 2002)*, pp. 1265–1270, 2002.

9. Nomura, T., Tasaki, T., Kanda, T., Shiomi, M., Ishiguro, H., and Hagita, N., Questionnaire-based research on opinions of visitors for communication robots at an exhibition in Japan, *International Conference on Human-Computer Interaction (Interact 2005)*, 2005.

10. Hayashi, K., Kanda, T., Miyashita, T., Ishiguro, H., and Hagita, N. Robot Manzai—Robots' conversation as a passive social medium, *IEEE International Conference on Humanoid Robots (Humanoids2005)*, 2005.

2.3 Humanoid Robots as a Passive-Social Medium— A Field Experiment at a Train Station

Kotaro Hayashi,[1,2] Daisuke Sakamoto,[3] Takayuki Kanda,[1] Masahiro Shiomi,[1,4] Satoshi Koizumi,[1,4] Hiroshi Ishiguro,[1,4] Tsukasa Ogasawara,[1,2] and Norihiro Hagita[1]

[1]*ATR Intelligent Robotics and Communication Labs.,*[2]*NAIST,*
[3]*The University of Tokyo,*[4]*Osaka University*

ABSTRACT

This section reports a method that uses humanoid robots as a communication medium. There are many interactive robots under development, but due to their limited perception, their interactivity is still far poorer than that of humans. Our approach in this paper is to limit robots' purpose to a non-interactive medium and to look for a way to attract people's interest in the information that robots convey. We propose using robots as a passive social medium, in which multiple robots converse with each other. We conducted a field experiment at a train station for eight days to investigate the effects of a passive social medium.

Keywords: Passive-social medium, human–robot interaction, communication robot, robot–robot communication

2.3.1 Introduction*

Over the past several years, many humanoid robots have been developed that can typically make sophisticated human-like expressions. We believe that humanoid robots will be suitable for communicating with humans, since their human-like bodies enable actual humans to intuitively understand their gestures, to the extent that humans sometimes unconsciously behave as if they were communicating with peers. In other words, if a humanoid robot effectively uses its body, people will communicate naturally with it. This could allow robots to perform communicative tasks in human society such as guiding people along a route [1]. Moreover, recent studies in embodied agents [2] and real robots [3] have revealed that robots can be used as a medium to convey information by using anthropomorphic expressions (Figure 2.7a). In addition, several advantages of robots over computer graphics agents have been demonstrated [3,4].

On the other hand, a robot's interaction capability is still far poorer than that of a human due to its limited sensing capability. This shortcoming is particularly noticeable when we introduce robots into our daily lives. Although the appearance of a humanoid robot often makes people believe that it is capable of human-like communication, it cannot currently engage in such sophisticated communication. At the forefront of robotics research remains the pursuit of sensing and recognition. For example, how can we make a robot recognize humans' postures and gestures or its environment? The results of such research should eventually be integrated into robots so that they can behave as ideal interaction partners that are capable of human-like communication. Pioneering research works in human–robot interaction (HRI) have revealed what robots can accomplish, such as museum guidance [5,6], perspective-taking [7], operation support [8], behaving as a well-mannered servant, and support for language study [9]; however, a robot's ability to inform humans is still quite limited.

Instead of struggling with the problem of insufficient interactivity built into a robot, we are exploring an alternative approach to maximizing the information that a robot system can offer to people, particularly focusing on attracting ordinary people's interest to the information. This new strategy is based on showing a conversation between robots. For example, Kanda et al. proved that users understand a robot's speech more easily and more actively respond to it after observing a conversation between two robots [10]. We named this kind of medium the "passive-social medium." Figure 2.8 illustrates the difference between this medium and other forms of human–robot interaction. At times, robots have been merely used for presenting information to people,

---

* This chapter is a modified version of a previously published paper Kotaro Hayashi, Daisuke Sakamoto, Takayuki Kanda, Masahiro Shiomi, Satoshi Koizumi Hiroshi Ishiguro, Tsukasa Ogasawara, and Norihiro Hagita, to be comprehensive and fit with the context of this book, *Humanoid robots as a passive-social medium—a field experiment at a train station*, ACM/IEEE *2nd Annual Conference on Human-Robot Interaction (HRI2007)*, pp. 137–144, 2007.

(a)

(b)

FIGURE 2.7
Scenes of the experiment: (a) robots as medium in station, (b) view of station.

which we call a passive medium (Figure 2.8a). This is the same as a news program on TV where an announcer reads the news. On the other hand, many researchers have been working to realize robots that act as an interactive medium (Figure 2.7c) that can accept requests from people as well as present information to people. However, due to difficulties with sensing and recognition, the resulting interactivity and naturalness is far from what people expect in a human-like robot.

The robot-conversation-type medium, on which we focus in this chapter, is named a passive-social medium (Figure 2.8b). It does not accept requests from people, as in the case of a passive medium, but attracts people's interest to information more than does a passive medium through its social

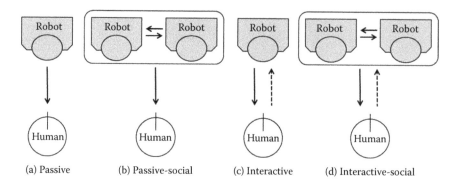

FIGURE 2.8
Robot(s) as medium.

ability, i.e., the expression of conversation. We believe that a "passive-social medium" is a more natural way to offer information to people than a simple passive medium. This is rather similar to a news program on television where announcers discuss comments told by others. Figure 2.8d shows what we call an *interactive-social medium,* but it has a weakness in its interactivity, as in the case of a conventional interactive robot medium.

To summarize these arguments, we are interested in two factors of robots as a medium: social expression (one robot or two robots) and response (whether it is interactive in a real field or not). By comparing the four robot-medium types (a)–(d), we investigate the optimal usage of robots as a medium.

Regarding the role of the robots, we assume that they will be used for advertisements and announcements. This is one of the simplest use of a medium. Moreover, since a robot seems novel enough to attract people's attention, we believe that this assumption is reasonable for earlier applications of robots. In order to study such robots as a medium, it is important to conduct a field experiment where ordinarily people pass by without having any motivation to interact with the robot, which is unlike having a subject come into a laboratory.

The opening of a new train station (Figure 2.7b) gave us a unique opportunity for real-world experimentation. We conducted a field experiment at the train station for eight days, where the robot was used to announce various features of the new train line. Through this experiment, we reveal the effects of a passive-social medium.

2.3.2 Multi-robot Communication System

The system consists of a sensor and humanoid robot(s) (Figure 2.9). The robots' behavior was controlled by a scenario-controlling system, which we expanded from our previous system [11] by adding a function that makes the robots change their behavior (thus, the scenario they are acting out) when

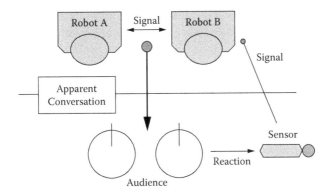

FIGURE 2.9
Outline of multi-robot communication.

a human is around. The robots' behavior is written in a simple scripting language that is easy to prepare.

2.3.2.1 Design Policy

The system implements social expression capabilities and interactivity that perform reliably in a real field. The social expression capability is based on a system we had already developed [11] that allows precise control of conversation timing and easy development. Regarding the interactivity, we limited it to be very simple but robust. The system immediately responds when a person comes close to the robot(s) by making the robot bow to the person. This limited-realistic interactivity is realized with a laser range-finder placed in front of the robot. We did not use any other sensors such as audition or vision, because outputs from such sensors are uncertain in a real field. Thus, what we refer to as "limited-realistic interactivity" is very different from that in some interactive robots, such as Robovie [6,9,12], where people may enjoy the unpredictability of unstable sensing capability. We decided on this implementation because unstable interactivity does not work when the purpose is to inform people. Users would be frustrated if they could not retrieve the information they needed.

2.3.2.2 Humanoid Robot

We used the humanoid robot Robovie [12] for this system. Figure 2.7a shows "Robovie," an interactive humanoid robot characterized by its human-like physical expressions and its various sensors. We used humanoid robots because a human-like body is useful in naturally catching people's attention. The human-like body consists of a body, a head, a pair of eyes, and two arms. When combined, these parts can generate the complex body movements required for communication. Its height is 120 cm and its diameter is

40 cm. The robot has two 4 × 2 degrees of freedom in its arms, three degrees of freedom in its head, and a mobile platform. It can synthesize and produce a voice through a speaker.

2.3.2.3 Sensor

To detect a human approaching a robot, we used a laser range-finder. The laser range-finder that we used is the LMS200 made by SICK. This sensor can scan 180° degrees horizontally, and it measures this range within a 5 m distance with a minimum resolution limit of 0.25° and 10 mm.

    The output is used to make the robots look at (turn their heads toward) the human passing by them. We simplified the output from the sensor by dividing the sensor's detection area into 19 areas (Figure 2.10) because 9° is a reasonable range to view when a robot looks at a walking person. The sensor sends a signal to the system according to the distance of the human, "1~19." If a human comes closer than a set distance, this sensor considers that the human is in front of the robot and sends signal "9." If there is no human in the area, this sensor sends signal "–1." In the case that more humans are in the observation range, only the closest is considered.

2.3.2.4 Scenario-Controlling System

This system is based on our scripting language for multi-robots [11], with a new function for changing the scenario that the robots act out when a human presence is detected. The scripting language has adequate capabilities for

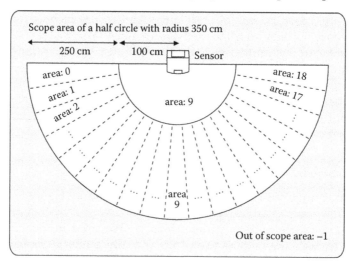

FIGURE 2.10
Sensing areas.

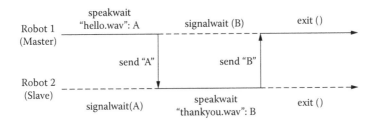

FIGURE 2.11
Example of signal exchanges.

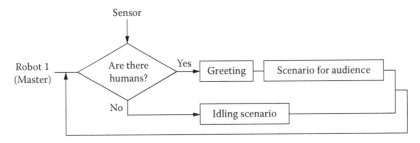

FIGURE 2.12
Example of restrictive reaction.

describing multi-robot communication and is simple enough for developer to easily use it to control the robots' behavior.

In this system, a set of robots interpret script files and execute scripts written in this language. One robot becomes the master. The master robot receives signals from the sensor, decides which scenario to execute, and informs its partner robot about it.

Figure 2.11 shows an example of the scripting language. In Figure 2.11, after Robot1 finishes playing the voice file "Hello.wav," it sends signal "A" to Robot2. At that time, Robot2 is waiting for signal "A." After receiving the signal "A," Robot2 plays the "thankyou.wav" file and sends signal "B." When Robot1 receives it, this scenario is finished.

Figure 2.12 shows how the system works in the interactive condition. If Robot1 (master robot) notices there is a person around, it decides which scenario to execute and sends the corresponding signal to its partner. When there is no human around, the robots play the idling scenario.

In this way, we realized a system that is capable of interpreting its environment and executing scenarios accordingly.

2.3.2.5 Example of a Script

Table 2.2 shows an example of a scenario that was actually used. In the interactive condition (see Section 2.3.3.3), when a human approached, the robot(s)

TABLE 2.2

Example of Scenario Announcing Station and Travel Information

---

Social Condition

Robot1: Thank you for using the train.

Robot2: Hey, you said this train can go to Osaka. Where can I go in Osaka, and how long does it take?

Robot1: It takes 30 minutes to Nanba and 40 minutes to Honmachi.

Non-social Condition

Robot1: Thank you for using the train.
I said this train can go to Osaka. Where in Osaka can I go, and how long does it take?
It is 30 minutes to Nanba and 40 minutes to Honmachi.

turned their faces toward the direction of the person and bowed while saying "Hello." After that, the robot(s) started to play the scenario (social or non-social, depending on the condition, see Section 2.3.3.3) for announcing station and travel information (Table 2.2).

2.3.3 Experiment

2.3.3.1 Method

Gakken Nara-Tomigaoka Station was opened in March 2006 as the terminal station of the Keihanna New Line, belonging to the Kintetsu Railway. The Keihanna New Line connects residential districts with the center of Osaka (Figure 2.7b). Station users are mainly commuters and students. There are usually four trains per hour, but in the morning and evening rush hours there are seven trains per hour. Figure 2.13 shows the experiment's environment. Most users go down the stairs from the platform after they exit a train.

We set the robots in front of the left stairway (Figure 2.13). Robots announced information toward users mainly coming from the left stairway. The contents are described in Section 2.3.3.4.

We observed how the users reacted to the behaviors of the robots. For observation, we set cameras on the ceiling nearby (Figure 2.13, cameras [a] and [b]).

Since there are four conditions (see Section 2.3.3.3), we prepared a time slot for each condition within a day. Moreover, since the number of users is different between the daytime and night, we divided the experiment time into four time slots that each cover both daytime (when there are mainly non-busy people) and night (when there are mainly busy commuters) to avoid incorrect results due to the difference in the number of users (Table 2.4).

2.3.3.2 Participants

All station users who passed by the robots were assumed to be participants. Their behavior was observed by video. We requested users who stopped to

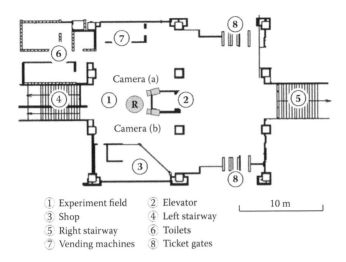

① Experiment field   ② Elevator
③ Shop   ④ Left stairway
⑤ Right stairway   ⑥ Toilets
⑦ Vending machines   ⑧ Ticket gates

FIGURE 2.13
Station map.

watch the robot(s) to answer a voluntary questionnaire. We obtained permission to record video from the responsible authorities of the station, and a notice was displayed in the station about the video recording.

2.3.3.3 Conditions

The following four conditions were prepared to investigate two factors: social expression and limited-realistic interactivity.

2.3.3.3.1 Passive (P) Condition

In this condition, one humanoid robot was installed (Figure 2.14a). The robot had a sensor in front of it, though the sensor was not used. The robot

(a)

(b)

FIGURE 2.14
Social expression: non-social condition (a) Social condition (b).

TABLE 2.3

Example of Idling Scenario

Social Condition
Robot1: *I'm hungry.*
Robot2: *Me, too. Let's eat a battery later.*
Non-social Condition
Robot1: *I'm hungry.*
I am going to eat a battery later.

continued to play the five scenarios (see Section 2.3.3.4) announcing station and travel information randomly.

2.3.3.3.2 Interactive (I) Condition

One humanoid robot was installed as in the P condition, but the robot had limited-realistic interactivity. That is, it had a sensor (laser range-finder) in front of it and changed the scenario according to the position of the human. Concretely, if there was no person near the robot, the robot played the idling scenario. In this scenario, robots talk to themselves (an example is shown in Table 2.3). When the sensor detected a person within a semicircle of 3.5 m, the robot stopped playing the idling scenario, looked at the person, bowed and said "Hello." After that, while one or more person were within the range of 3.5 m, the robot started to play the five scenarios announcing station and travel information randomly.

2.3.3.3.3 Passive-Social (Ps) Condition

Two humanoid robots were installed (Figure 2.14b). The robots had a sensor in front of them, but the sensor was not used. The robots continued to play the five scenarios announcing station and travel information randomly by communicating with each other.

2.3.3.3.4 Interactive-Social (Is) Condition

Two humanoid robots were installed as in the Ps condition. The robots had limited-realistic interactivity: the robots had an operating sensor in front of them and changed the scenario according to the position of the human. Concretely, if there was no person near the robots, the robots played the idling scenario. (In this scenario, robots chat with each other.) When the sensor detected a person within a 3.5-m-radius semicircle, the robots stopped playing the idling scenario, looked in the direction of the person, bowed and said "Hello." After that, when one or more person was within a range of 3.5 m, robots started to play the scenario of announcing station and travel information by communicating with one another.

2.3.3.4 Content of Scenarios

In each condition, there were five scenarios announcing station and travel information as follows:

1. Travel duration to Osaka
2. Information about ATR
3. Where passengers can go on this train line
4. Information about east Osaka (connected by the new line)
5. Facilities near the station.

These scenarios lasted about 3 minutes each on average. Table 2.2 shows a part of the scenario announcing travel duration to Osaka, which is shortened by the new line. In the interactive condition, there were five idling scenarios in addition to the information scenario. Table 2.3 shows one of the scenarios. These idling scenarios lasted about 30 s on average. After the fourth day, we changed the contents of these information scenarios.

2.3.3.5 Measurement

2.3.3.5.1 Questionnaire

We requested station users who stopped to watch the robots to answer a questionnaire. We obtained answers from 163 station users. The questionnaire had three questions as follows in which they rated items on a scale of 1 to 7, where 7 is the most positive:

Feeling of being addressed by the robot

Interest in the content of the information the robots is announcing

Enjoyment

2.3.3.5.2 Analysis of Behaviors

We analyzed all videos from cameras and recorded during the experiment period (Table 2.4) of the eight days of the experiment. In total, about

TABLE 2.4

Schedule for Robot Observation

Day (per hour)	Night (per 30 minutes)
14:00–15:00	18:30–19:00
15:00–16:00	19:00–19:30
16:00–17:00	19:30–20:00
17:00–18:00	20:00–20:30

FIGURE 2.15
Station users' behaviors toward the robot(s): (a) Ignoring, (b) noticing, (c) stopping to watch, (e) touching, (f) changing course to investigate, (g) talking about robot(s), (h) watching with child, (i) taking pictures.

5,900 people were observed. As a result, we found the following types of people's reactions to the robots.

(a) Ignoring   People passed by without noticing the robots (Figure 2.15a); 2,964 people showed this behavior. This case was noticed about 370 times a day.

(b) Noticing   People passed near the robots and saw the robots but did not stop walking (Figure 2.15b); 2,039 people showed this behavior. This case was noticed about 260 times a day.

(e)

(f-1)

(f-2)

(g)

(h)

(i)

FIGURE 2.15 (continued)

(c) Stopping to Watch People passed near the robots and stopped to watch the robots (Figure 2.15c). Usually, after they finished listening to one or two of the information items that the robots announced, they left. This case was noticed about 110 times a day.

Note that (c)–(i) are not mutually exclusive and (c) is inclusive of (d)–(i). For later analysis, we only used (c).

(d) Staying This case is a type of "stopping to watch," where people stopped to watch the robots and continued listening to the information that the robots announced for an extended time. There were five kinds of

information, and the robots repeated these randomly; therefore, the same information appeared many times during their stay. However, they kept watching the robots without getting tired. This case was noticed about once a day.

(e) Touching   In this case, people touched the robot as soon as they approached. In particular, two cases were found most frequently. In the first case, they touched the shoulder of the robot and confirmed its feel. In the second case, they held an arm as a robot raised it while motioning to explain information (Figure 2.15e). In some rare cases, we found that people pulled up the arm of a robot and hugged a robot without thinking that the robot might break. This case was noticed about three times a day.

(f) Changing Course to Investigate   In this case, people changed their course to come near the robots to investigate (Figure 2.15f). This case was often noticed about 50 times a day.

(g) Talking about Robot(s)   In this case, some people talked to each other about the robots. Usually, they talked with their friends, but in rare cases, some people talked with others whom they appeared not to know (Figure 2.15g). In this figure, they talked about the fact that the robots had a conversation with each other. This case was noticed about seven times a day.

(h) Watching with Child   In this case, people watched robots with a child. This case was noticed about 15 times a day. We show some examples as follows.

- A child found the robots and came to watch while pulling along his or her parents.
- A person found the robots and called his or her child to watch.

Usually, parents and children watched the robots eagerly. Some of the children sat down to watch (Figure 2.15h). However, we found some cases where the parents turned their attention in another direction as they lost interest in the robots in contradiction to the eagerness of the children. On the contrary, we also found some parents who watched the robots eagerly while the children turned their attention in another direction as they lost interest in the robots.

(i) Taking Pictures   In this case, people took pictures with a camera or a mobile phone. For some of them, it seemed more important to take pictures than to listen to the information that the robots announced. This case was noticed about seven times a day (Figure 2.15i).

We found these typical nine cases as described above. Then, we summarized cases (d)–(i) into "(c) Stopping to Watch" because these cases were not mutually exclusive and numbered too few for statistical analysis. As a result, about 900 people were categorized in the "Stopping to Watch" category.

2.3.3.6 Hypotheses

We examined the following hypotheses in this experiment.

Hypothesis 1
If the robot(s) is "interactive" with people by bowing to them before announcing the information, the people will get a stronger feeling of being addressed by the robot.

Hypothesis 2
If a robot is passive toward people without reacting to them, people will pay more attention to the information coming from the robot(s).

Hypothesis 3
People are more likely to stop to listen to the robot's conversation in a two-robot condition than in a one-robot condition.

2.3.3.7 Results

2.3.3.7.1 Verification of Hypothesis 1: Feeling of Being Addressed

Figure 2.16 shows the results of "feeling of being addressed by the robot." A two-way (sociality × interactivity) between-group ANOVA (analysis of variance) was conducted, which showed a significant difference between the interactive condition and the passive condition. The interactive condition is

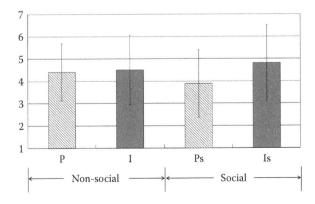

FIGURE 2.16
Feeling of being addressed by the robot.

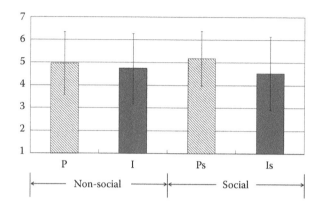

FIGURE 2.17
Interest in information robot(s) announcing.

higher than the passive condition ($F(1,136) = 4.63$, $p < .05$), but there is no significant difference in the sociality factor ($F(1,136) = .18$, $p > .10$) and an almost significant effect in interaction ($F(1,136) = 2.97$, $p = .087$). That is, the interactivity of the robot gives people a stronger feeling of being addressed by the robot.

2.3.3.7.2 Verification of Hypothesis 2: Interest in Information

Figure 2.17 shows the results of "interest in information." As the result of a two-way between-group ANOVA, the passive condition is significantly higher than the interactive condition ($F(1,136) = 4.11$, $p < .05$). There is no significant difference in the sociality factor ($F(1,136) = .00$, $p > .10$) or interaction ($F(1,136) = .97$, $p > .10$). That is, the limited-realistic interactivity of the robots makes people more likely to lose interest in the information (see Figure 2.18).

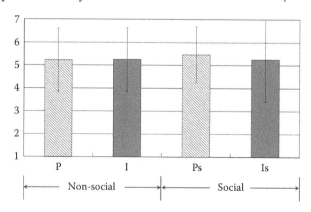

FIGURE 2.18
Enjoyment.

TABLE 2.5

Data of Stopping Conditions

Condition	Number of Stopping to Watch	Number of Ignoring or Noticing	Rates of Stopping to Watch
P	197	1278	0.134
I	214	1330	0.139
Ps	242	1264	0.161
Is	246	1131	0.179

A = Number of persons stopping to watch, B = Number of persons ignoring or noticing, C = Rate of persons stopping to watch, A + B = Number of All persons, C = A/[A + B].

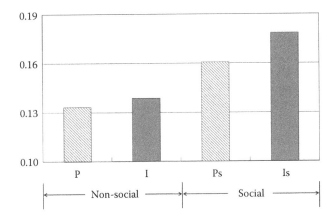

FIGURE 2.19
Rate of "stopping to watch."

2.3.3.7.3 Verification of Hypothesis 3: Rate of Stopping to Watch

Table 2.5 and Figure 2.19 shows the results of the number of people "stopping to watch" for each condition. From the results of the chi-square test, we see there are significant differences between the conditions. The "interactive-social" condition is significantly high, while nonsocial conditions ("interactive" and "passive") are significantly low. That is, people are more likely to stop to listen to the conversation of two robots.

2.3.3.7.4 Analysis of Enjoyment

As a result of a two-way between-group ANOVA, there is no significant difference in either the sociality factor ($F(1,136) = .27, p > .10$), the interactivity factor ($F(1,136) = .18, p > .10$), or interaction ($F(1,136) = .30, p > .10$). (Figure 2.18).

2.3.3.7.4.1 Summary of Results The results indicate that limited-realistic interactivity of the robot gives people the feeling of being addressed by the robots. On the other hand, it makes people lose interest in the information. From this result, we believe that using "interactive" as a medium does not necessarily provide a good result in its current form, since such performance has limited realistic use in a real field. Moreover, the nonsocial conditions had a lower chance of making people stop at the robot. These findings indicate that the passive-social medium is promising because the system has a better chance of getting people to stop and become interested in the information announced by the robot.

2.3.4 Overview

2.3.4.1 Contribution to HRI Research

As one of its major contributions, this research demonstrated the positive potential of communication robots at a train station: people at the station sometimes stopped at the robots and listened to the robots' speech. Important information was also reported because people's reactions were observed at the station. Previous research focused on museums [5,6], universities [13], and schools [9] where people have a tendency to be interested in robots. Therefore, it was not clear how the public in general would react to such a human-like robot that talks to people. Moreover, in a train station there are many people who do not want to interact with the robots and who are typically busy with their travels. We believe it is worthwhile to conduct a field experiment at such a busy place as well as places where people are highly interested in robots.

Mainly nonbusy people seemed to stop walking to interact with the robots as we expected, and the majority of the people did not even look at the robot as they passed through the station. On the other hand, several people were very interested in the robots, performing such actions as talking about the robots while looking at them, touching and talking to them, and looking at them for a long time, which has also been observed in different field trials at places such as a science museum.

2.3.4.2 Contribution to Robot Design as a Medium

This research showed how a robot as a medium at a public place, such as a station, should be designed. The experiment revealed that low-level interactivity by the robot increased the feeling of actually conversing with the robot but decreased people's interest in the contents of the messages. That is, low-level interactivity can introduce an obstruction for the use of robots as a medium for conveying information to people.

Regarding the interactivity of robots, researchers are examining ways to improve this. Robots are capable of finding human faces, postures, gestures, and so forth within laboratories, but a real field situation such as a train station is still a difficult environment for using such sensing capabilities. This research demonstrated one alternative approach for using robots in a real field site.

2.3.4.3 Effects of Robots as Passive-Social Medium

Although we have used robots as a passive-social medium at a science museum [6] and a *manzai* performance [11], we have not revealed the effects of robots as a passive-social medium in comparison with other forms. In both trials, robots got people's attention so that they crowded around to see the robots. One of the difficulties has been that when people have a strong interest in robots, it is difficult to identify the effects of the passive-social medium because people appreciate an encounter with any kind of robot due to its novelty.

The experimental results revealed that a two-robot condition (passive-social and interactive-social conditions) was better than a one-robot condition in terms of getting people to stop at the robots. Once people stopped, these conditions did not make any difference. Instead, a lack of interactivity (passive-social and passive conditions) produced the advantage of attracting people's interest in the contents of the utterances. Thus, the passive-social condition proved to be best for this purpose among the conditions tested in the experiment.

Although the experiment revealed the positive aspect of passive-social medium on the "interest" aspect, it is not clear how naturally the passive-social medium offers information compared with other types of medium. The experimental results revealed effects when people glanced at the robot to decide whether to stop; however, the results did not reveal effects after stopping at the robots. The difficulty is in experimental control. In this experiment, we controlled the contents that the robots said. Two robots (passive-social condition) enable us to play a bigger variety of scenarios than it is possible with a single robot. For example, one robot might ask a question to another, after which the other would make a response. Use of such a stage effect, however, would cause differences not only due to the conditions (passive-social versus passive) but also due to the different contents of utterances. Thus, we did not implement such techniques in this experiment. Probably, adding such a feature would make robots more enjoyable and make interaction with people more natural. Demonstrating such effects will be one of our future studies.

2.3.4.4 Novelty Effect

Previous research reported a novelty effect, which is the phenomenon of people rushing to interact with a robot at the beginning and then rapidly losing their passion to interact with it [9]. A similar phenomenon might be expected at a train station because there are many people using the station daily to commute to their offices and schools. However, this was not observed. Not so many people gathered around the robot, and the frequency of stopping

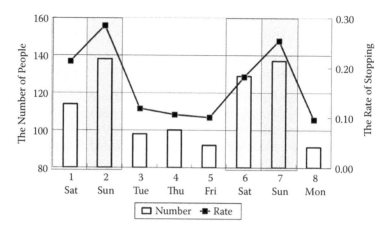

FIGURE 2.20
Number and rate of "stopping to watch" on each day.

at the robot did not decrease (Figure 2.20). Perhaps, then, robots are not such novel objects for the general public at a station.

2.3.4.5 Limitations

Since this research was conducted with only one particular robot, Robovie, with a small number of utterances, and at a train station in a residential area, the generalities of robots, stations, and contents are limited.

Regarding the generality of robots, a previous research work demonstrated little difference in people's responses to different robots [14], so we expect that a similar response can be obtained with different humanoid robots. Regarding the generality of stations, people's reactions will probably be different at a busy and crowded station in an urban area; however, it would be difficult to conduct such an experiment at a busy and crowded station due to safety problems. As for the generality of contents, we believe that the setting was realistic for conducting an experiment. We tried to make the contents as simple as possible to reveal basic differences among conditions. When this kind of robot is used for real applications such as advertisements and announcements, we expect that the contents will be more sophisticated, including the use of humor, which could more effectively elicit people's reactions and possibly weaken the differences among conditions.

2.3.5 Conclusion

A field experiment was conducted at a train station for 8 days, using the technology developed for a multi-robot communication system. The robots' task was to inform passengers of station and travel information.

The purpose of the experiment was to identify the best way of informing users. The effects of two factors, social expression and limited-realistic interactivity, were studied. The results indicate that a passive-social medium (social but not interactive) was the most effective way of attracting people's interest in the information; the interactivity was useful in giving people the feeling of talking with the robots. This implies that we should use different forms of robots according to the purpose: a passive-social medium for advertising and an interactive medium for peer-to-peer conversation, such as guiding along a route and exchanging detailed information adapted for each individual.

Acknowledgments

We wish to thank the staff of the Kinki Nippon Railway Co. Ltd. for their kind cooperation. This research was supported by the Ministry of Internal Affairs and Communications of Japan.

References

1. D. Sakamoto, T. Kanda, T. Ono, M. Kamashima, M. Imai, and H. Ishiguro, Cooperative embodied communication emerged by interactive humanoid robots, *International Journal of Human-Computer Studies*, Vol. 62, pp. 247–265, 2005.
2. J. Cassell, T. Bickmore, M. Billinghurst, L. Campbell, K. Chang, H. Vilhjalmsson, and H. Yan, Embodiment in conversational interfaces: Rea, *Conf. on Human Factors in Computing Systems (CHI'99)*, pp. 520–527, 1999.
3. C. Kidd and C. Breazeal, Effect of a robot on user perceptions, *Int. Conf. on Intelligent Robots and Systems (IROS'04)*, 2004.
4. K. Shinozawa, F. Naya, J. Yamato, and K. Kogure, Differences in effect of robot and screen agent recommendations on human decision-making, *International Journal of Human-Computer Studies*, Vol. 62, pp. 267–279, 2005.
5. R. Siegwart et al., Robox at Expo.02: A large scale installation of personal robots, *Robotics and Autonomous Systems*, Vol. 42, pp. 203–222, 2003
6. M. Shiomi, T. Kanda, H. Ishiguro, N. Hagita, Interactive humanoid robots for a science museum, *ACM 1st Annual Conference on Human-Robot Interaction (HRI2006)*, pp. 305–312, 2006.
7. G. Trafton, A. Schultz, D. Perznowski, M. Bugajska, W. Adams, N. Cassimatis, D. Brock, Children and robots learning to play hide and seek, *ACM 1st Annual Conference on Human-Robot Interaction (HRI2006)*, pp. 242–249, 2006.

8. C. Breazeal, C. Kidd, A. L. Thomaz, G. Hoffman, M. Berlin, Effects of nonverbal communication on efficiency and robustness in human-robot teamwork, *Proceedings of IEEE/RSJ International Conference on Intelligent Robotis and Systems (IROS 2005)*, 2005.

9. T. Kanda, T. Hirano, D. Eaton, and H. Ishiguro, Interactive robots as social partners and peer tutors for children: A field trial, *Human Computer Interaction*, Vol. 19, No. 1–2, pp. 61–84, 2004.

10. T. Kanda, H. Ishiguro, T. Ono, M. Imai, and K. Mase, Multi-robot cooperation for human-robot communication, *IEEE International Workshop on Robot and Human Communication (ROMAN2002)*, pp. 271–276, 2002.

11. K. Hayashi, T. Kanda, T. Miyashita, H. Ishiguro, N. Hagita, Robot Manzai—Robots' conversation as a passive social medium-, *IEEE International Conference on Humanoid Robots (Humanoids2005)*, pp. 456–462, 2005.

12. T. Kanda, H. Ishiguro, M. Imai, and T. Ono, Development and evaluation of interactive humanoid robots, *Proceedings of the IEEE*, Vol. 92, No. 11, pp. 1839–1850, 2004.

13. R. Gockley, J. Forlizzi, and R. Simmons, Interactions with a moody robot, *ACM 1st Annual Conference on Human-Robot Interaction (HRI2006)*, pp. 186–193, 2006.

14. T. Kanda, T. Miyashita, T. Osada, Y. Haikawa, and H. Ishiguro, Analysis of humanoid appearances in human-robot interaction, *IEEE/RSJ International Conference on Intelligent Robots and Systems (IROS2005)*, pp. 62–69, 2005.

2.4 An Affective Guide Robot in a Shopping Mall

Takayuki Kanda, Masahiro Shiomi, Zenta Miyashita,
Hiroshi Ishiguro, and Norihiro Hagita

                          ATR Intelligent Robotics and Communication Laboratory

ABSTRACT

To explore possible robot tasks in daily life, we developed a guide robot for a shopping mall and conducted a field trial with it. The robot was designed to interact naturally with customers and to affectively provide shopping information. It was also designed to repeatedly interact with people to build a rapport; since a shopping mall is a place people repeatedly visit, it provides the chance to explicitly design a robot for multiple interactions. For this capability, we used RFID tags for person identification. The robot was semi-autonomous, partially controlled by a human operator, to cope with the difficulty of speech recognition in a real environment and to handle unexpected situations.

A field trial was conducted at a shopping mall for 25 days to observe how the robot performed this task and how people interacted with it. The robot interacted with approximately 100 groups of customers each day. We invited customers to sign up for RFID tags and those who participated answered questionnaires. The results revealed that 63 out of 235 people in fact went shopping based on the information provided by the robot. The experimental results suggest promising potential for robots working in shopping malls.

Keywords: Communication robots, service robots, field trial

2.4.1 Introduction[*]

Recent progress in robotics has enabled us to start developing humanoid robots that interact with people and support their daily activities. We believe that humanoid robots will be suitable for communicating with humans. Previous studies have demonstrated the merits of the physical presence of robots for providing information [1,2]. Moreover, their human-like bodies enable them to perform natural gaze motion [3] and deictic gestures [4].

These features of humanoid robots might allow them to perform such communicative tasks in human society as route guidance and to explain exhibits. Since we are not yet sure what the communicative tasks of robots will be like, many researchers have started to conduct field trials to explore possible tasks. Such explorations are one important research activity, since HRI remains a very young field where few real social robots are working in daily environments. Field trials enable us to envision the future scenes of human–robot interaction and their accompanying problems that must be solved so that robots can be accepted in daily life.

Previous studies have revealed that social robots can be used as museum guides [5, 6], as a receptionist for assisting visitors [26], as peer-tutors in schools [7], in the context of mental-care for elderly people [8], in autism therapy [9,10], and child-care [11]. In contrast, our study focuses on an information-providing task in daily environments at a shopping mall. Compared with schools and museums, a shopping mall is a public environment open to ordinary people who are often busy, who are not seeking a tool to "kill" time, and who do not have special interest in robotics technology: the environment is challenging. This paper aims to answer the following questions:

Can a robot function in an information-providing task in an open public environment, such as a shopping mall?

Can a robot influence people's daily activities, such as shopping?

Can a robot elicit spontaneous repeated interaction?

[*] This chapter is a modified version of a previously published paper Takayuki Kanda, Masahiro Shiomi, Zenta Miyashita, Hiroshi Ishiguro, and Norihiro Hagita, An affective guide robot in a shopping mall, *4th ACM/IEEE International Conference on Human-Robot Interaction(HRI2009)*, pp. 173–180, 2009, edited to be comprehensive and fit with the context of this book.

In the chapter, we report a number of technical challenges for a robot in a shopping mall. One notable feature is using a human operator to cope with difficulties in real environments. This is quite often used in human–computer interaction (HCI) and HRI for prototyping, known as Wizard of Oz (WOZ) [12,13]. In addition, since our vision is to use a human operator for more than making a prototype, we believe that a robot could start working in daily environments with human operators with a technique that minimizes the task load of operators, such as one with which one operator can control four robots [14].

2.4.2 Design

There are two aspects of design related to HRI: one is the appearance that considers impressions and expectations [15,16]. Another aspect concerns behavior design, e.g., a scenario design for assisted play [17] and design patterns of interactive behaviors [4].

These two aspects are, of course, mutually related; however, in this study we focused on the latter direction to establish how we can design a robot's interactive behavior for a shopping mall. Here, we introduce how we designed the robot's roles, how we realized them in the system framework, and how we considered them while creating the robot's interactive behaviors. We believe that this provides a chance to start considering the design process of such a social robot that works in the real world.

2.4.3 Contemplating Robot Roles

What kind of robots do people want in their daily lives? According to a Japanese government report [18], a majority of respondents believe that providing information at such public spaces as stations and shopping malls is one desired task of robots.* People also want machines like robots to perform physical tasks, such as toting luggage. Thus, we decided to explore an information-providing task for a robot in a public space as a guide robot at a mall with many shops nearby.

The next question addresses the roles of a guide robot in a mall. Many other facilities, such as maps and large screens, provide information. In contrast, a robot has unique features based on its physical existence, its interactivity, and its capability for personal communication. We defined the three roles based on this consideration.

---

* This might be relatively high in Japan rather than other countries: 76.2% of the respondents think it is good to have robots at a transportation facility such as a station and 87.5% of them think at-the-place guidance is a good task for robots; 64.2% of the respondents think it is good to have robots at a commercial place, such as a shopping mall, and 87.9% of them think at the place guidance is a good task for robots.

2.4.3.1 Role 1: Guiding

The size of shopping malls continues to become larger and larger. Sometimes people get lost in a mall and ask for directions. Even though a mall has maps, many people still prefer to ask for help. Some information is not shown on a map; thus, people ask "Where can I buy an umbrella?" or "Where can I print a digital camera?" (The author was actually asked this strange question in a mall, which seems to suggest the need of human support for the robot.) Here, a route guidance service is needed.

In contrast to a map or other facilities, a robot has unique features: a physical existence, it is co-located with people, and it is equipped with human-like body properties. Thus, as shown in Figure 2.21, a robot can naturally explain a route like humans by pointing to it, looking in the same direction as the person is looking, and using such reference terms as "This way."

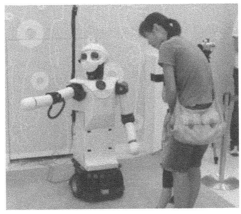

FIGURE 2.21
Robot guiding a customer with deictic representation.

2.4.3.2 Role 2: Building Rapport

From the customer view, since a robot is one representative of the mall, it needs to be friendly so that customers feel comfortable. In addition, since a mall is a place that people repeatedly visit, a robot needs to naturally repeat interaction with the same person; thus, a function that builds rapport with each customer is useful. The importance of building rapport has been studied in HCI in the context of affective computing [19].

Moreover, one future scenario in this direction is a function of customer relationship management. Previously, this was done by humans: for example, in a small shop, the shopkeeper remembers the "regulars" and molds communication to each individual. For example, he/she might be particularly cordial to the good customers who often frequent the shop. Recently, since the number of customers is too unwieldy to manage, information systems have assumed this role in part, such as the mileage services of airplane companies, the point systems of credit cards, and online shopping services such as Amazon. However, these information systems do not provide natural personalized communication as humans do; in contrast, we believe that in the future a robot might be able to provide natural communication and personalized service for individual customers and develop relationships or a rapport with them.

2.4.3.3 Role 3: Advertisements

From the mall's point of view, advertising is one important device or facility they need. For instance, posters and signs are placed everywhere in malls. Recently, information technologies are being used for such purposes as well. Figure 2.22 shows a large screen (about 5 m by 2.5 m) for providing shopping information to customers, placed in the shopping mall where we conducted our field trial. The screen shows such shop information as places in the mall, product features of the shops, etc.

We believe that a robot can also be a powerful tool for this purpose. Since a robot's presence is novel, it can attract people's attention and redirect their interest to the information it provides [20]. In addition, it can provide information to people in a way people talk together; for example, it can mention shops and products from its first-person view (See 2.3.4).

2.4.4 System Design

The robot's role is limited by its recognition and action capabilities, which are largely limited by its hardware and infrastructure. Thus, first, we should consider system design (hardware and infrastructure). In HRI, we need to explore a promising combination of hardware and infrastructure. Some researchers are studying a stand-alone robot that has all sensing, decision-making, and acting capabilities. In contrast, some researchers are focusing on a combination of robots, ubiquitous sensors, and humans.

FIGURE 2.22
Shopping mall and its large information screen.

We have chosen the latter strategy, known as a "network robot system" [21], in which a robot's sensing is supported by ubiquitous sensors and its decision processes by a human operator.

From a user view, the central component is a robot that provides information in a natural way with its speaking capability as well as its body properties for making gestures. Thus, regardless whether it is a stand-alone or a networked robot system, users can concentrate on the robot in front of them.

In contrast to the user view, in a network robot system, most of the intelligent processing is done apart from the robot. Sensing is mainly done by ubiquitous sensors. There are three important sensing elements in our system: *position estimation, person identification,* and *speech recognition.* For *position estimation,* we used floor sensors that accurately and simultaneously identify the positions of multiple people. This could also be done with other techniques, such as a distance sensor. For *person identification,* we employed a passive-type radio frequency identification (RFID) tag that always provides accurate identification. Such tags require intentional user contact with a RFID reader; since passive-type RFIDs have been widely adopted for train tickets in Japan, we consider this unproblematic.

We used a human operator for *speech recognition* and *decision making.* For this way of providing information, instability and awkwardness would cause critical disappointment, and the quality of current speech recognition technology remains far from useful. For instance, a speech recognition system prepared for noisy environment, which performs 92.5% word accuracy in 75 dBA noise [22], resulted in only 21.3% accuracy in a real environment [23]. This reflects the natural way of daily utterances, the changes of voice volume among people and/or within the same person, and the unpredictability of noise in a real environment. Thus, since a speech recognition program causes many recognition errors, the robots have to ask for elucidation too often.

2.4.5 Behavior Design

2.4.5.1 General Design

We set two basic policies for designing the robot's interaction. First, it takes the communication initiative and introduces itself as a guide robot. It asks about places and then provides information in response to user requests. Thus, customers clearly understand that the robot is engaged in route guidance.

Second, its way of utterance and other behaviors are prepared in an affective manner [19], not in a reactive manner. The robot engages in human-like greetings, report its "experience" with products in shops and tries to establish a relationship (or rapport) [24] with the individuals. This is very different from master–slave-type communication where a robot prompts a user to provide a command.

2.4.5.2 Guiding Behavior

There are two types of behaviors prepared for guiding: *route guidance* and *recommendation.* The former is a behavior in which the robot explains a route to a destination with utterances and gestures, as shown in Figure 2.21. The robot points in the first direction and says, "Please go that way" with an appropriate reference term chosen by an attention-drawing model [25]. It continues the explanation, saying: "After that, you will see the shop on your right." Since the robot knows all of the mall's shops and facilities (toilets, exits, parking, etc.), it can explain 134 destinations.

In addition, for situations where a user hasn't decided where to go, we designed *recommendation* behaviors in which the robot suggests restaurants and shops. For example, when a user asks, "Where is a good restaurant?" the robot starts a dialogue by asking a few questions, such as "What kind of food would you like?" and chooses a restaurant to recommend.

2.4.5.3 Building Rapport Behavior

For persons wearing RFID tags, the robot starts to build rapport through a dialogue that consists of the following three policies:

Self-disclosure: The importance of self-disclosure for humans to identify friendly behavior has long been studied. Bickmore and Picard used this strategy as a relational agent for building relationships with users [24]. Gockley et al. made a receptionist robot that tells new stories and successfully attracted people to interact with it [26]. In our previous study, which was successful, our robot disclosed a secret [28]. In this study, we follow the same strategy: letting the robot perform self-disclosure. For example, the robot mentions its favorite food, "I like *takoyaki*," and its experiences, such as, "This is my second day working in this mall."

Explicit indication of person being identified: Since we found that people appreciated having their names used by robots in our previous studies [11], we continued this strategy. The robot greets a person by the name under which he/she registered, such as "Hello, Mr. Yamada." In addition, it uses a history of previous dialogue to inform that the robot remembers the person. For example, on day one, if the robot asked, "Do you like ice cream?" and if the person answered, "Yes," the robot says, "Ok, I'll remember that"; on day two, the robot says, "I remember that you said you like ice cream, so today, I'm going to tell you my favorite flavor of ice cream."

Change of friendliness in behaviors: For a person who repeatedly visits, the robot gradually changes its behavior to show a more and more friendly attitude. For example, on day one, it says, "I'm a little nervous talking with you for the first time"; but on day three it says, "I think we are friends" to show its warm attitude toward the person.

2.4.5.4 Behavior for Advertisements

The robot is also intended to provide advertisements about shops and products in a manner that resembles *word of mouth*. When the robot starts a conversation with a customer, it starts with a greeting and then engages in *word of mouth* behavior as a form of casual chat. It affectively reports its pretended experiences about products in shops. For example, the robot might say,

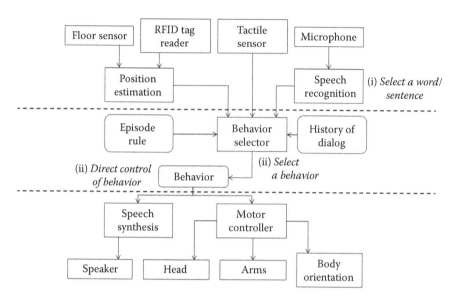

FIGURE 2.23
System configuration. In the figure, italic text represents operator role.

"Yesterday, I ate a crêpe in the food court. It was nice and very moist. I was surprised!" "The beef stew *omurice* at Bombardier Jr. was good and spicy. The egg was really soft, too, which was also very good." We implemented five topics per day and changed the topics every day so that daily shoppers didn't get bored with this behavior.

2.4.6 System Configuration

Figure 2.23 shows an overview of the system configuration. The robot identifies a person with an RFID tag reader and continues to track his/her position with floor sensors. As a WOZ method, speech recognition is conducted by a human operator. This information is sent to a behavior selector, which chooses an interactive behavior based on preimplemented rules called *episode rules* and the history of previous dialogues with this person.

2.4.7 Autonomous System

2.4.7.1 Robovie

"Robovie" is an interactive humanoid robot characterized by its human-like physical expressions and its various sensors (Figure 2.21). We used humanoid robots because a human-like body is useful for naturally capturing and holding the attention of humans [11]. It is 120 cm high, 40 cm wide, and has tactile sensor elements embedded in the soft "skin" that covers its body.

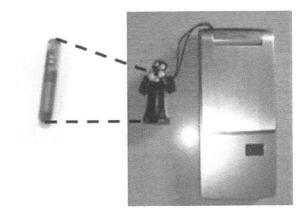

(a)

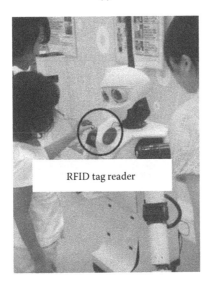

RFID tag reader

(b)

FIGURE 2.24
RFID tag and reader.

2.4.7.2 Person Identification

We invited customers at the field trial to register for an RFID tag for person identification. Figure 2.24a shows a passive-type RFID tag embedded in a cellular phone strap. The accessory is 4 cm high. The RFID tag's reader is attached to the robot's chest. Since a passive-type RFID system requires contact distance for reading, users were instructed to place the tag on the tag reader for identification and to interact with the robot (Figure 2.24b).

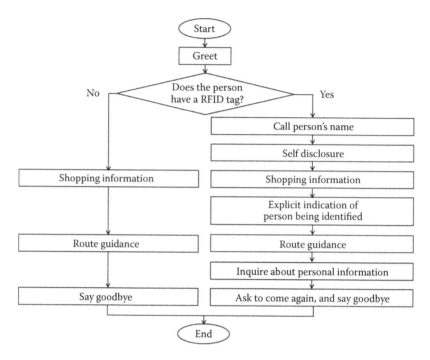

FIGURE 2.25
Flow of robot's dialogue.

2.4.7.3 Position Estimation

We installed 16 floor sensor units around the robot that covered a 2 × 2 m area (Figure 2.26). Each sensor unit is 50 × 50 cm with 25 on/off pressure switches. The floor sensors have a sampling frequency of 5 Hz. To estimate people's positions, we used a Markov Chain Monte Carlo method and a bipedal model [27]. This method provided robust estimation of positions without occlusion; the average position error was 21 cm. Thus, it was useful for situations where a person closely interacted with the robot.

2.4.7.4 Behavior and Episode Rules

Interactive behavior is implemented with *situated modules* ("behavior" in this paper) and episode rules [28]. Each situated module controls the robot's utterances, gestures, and nonverbal behaviors in reaction to a person's action. For example, when it offers to shake hands by saying, "Let's shake hands," it waits for input from a tactile sensor to react to the person's handshake. Each behavior usually lasts about 5 to 15 seconds. There were 1759 behaviors implemented, based on our design policies.

Episode rules describe the transition rules among the situated modules. There were 1015 episode rules implemented. An overview of the interaction

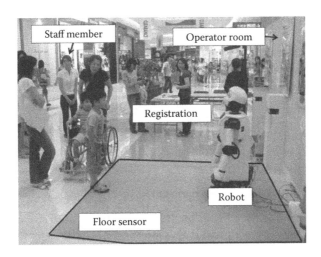

FIGURE 2.26
Field trial environment.

flow is summarized in Figure 2.25. When the robot detects a person, it greets that person. If the person touches his/her RFID tag, the robot starts the flow in the first branch of Figure 2.25. It calls the person's name, provides shopping information of the day, chats about the person's preferences, and offers route guidance. If the person does not have an RFID tag, it engages in simpler interaction providing shopping information and route guidance.

2.4.7.5 Nonverbal Behaviors

In addition to the interactive behaviors implemented as situated modules, the robot is also designed to sustain interaction nonverbally. The robot orients its body direction to the interacting person whose x-y position is detected with the floor sensor (explained in Section 2.4.7.3). Moreover, we implemented gaze behavior. The robot looks at a face of an interacting person; for this purpose, we inputted person's height information in the robot associated with ids of RFID tags, so that the robot is able to orient its gazing direction to the face. During guiding behavior, it points and looks at the direction (Figure 2.21) for shared attention.

2.4.8 Operator's Roles

The robot system is designed to operate without an operator; however, since speech recognition often fails in a noisy environment, there are many unexpected situations in dialogues where the robot cannot correctly respond to requests. Thus, in this study, we used the WOZ method with a human operator to supplement these weaknesses of autonomous robots. The detailed roles are described in the following subsections.

We made an important principle for the operator. In principle, we asked the operator to minimize the amount of operations. This principle is for studying the potential of robot autonomy. Except for substituting speech recognition, the operator only helped the robot when the operation was truly needed. For example, even if a user interrupted the flow and asked, "How old are you?" (a frequently asked question), the operator did not operate the robot. If the user continued to repeat the question without showing signs of stopping, the operator selected the robot's behavior, or even typed its utterance to answer.

2.4.8.1 Substitute of Speech Recognition

When a robot performs a behavior in which it asks a user a question, the teleoperation system prompts the operator to choose the words from the list expected for this situation. For example, when the robot asks, "I can give you the route. Where would you like to go?" the teleoperation system shows a list of places. When the robot asks, "Do you like ice cream?" it shows a simple choice of "yes," "no," and "no response" to the operator. Here, the operator behaves in the same way as speech recognition software. After the operator chooses the words, the robot autonomously continues the dialogue with the user.

2.4.8.2 Supervisor of Behavior Selector

There are significant degrees of uncertainty about user behavior toward the robot. Sometimes people asked about unexpected things. Even though the robot has a behavior to answer the question, here, the problem is the lack of episode rules. For example, although the robot has behaviors to guide and explain all of the shoe stores, it was confused when a user asked about a "shop for children's shoes," which was not in the speech recognition dictionary. For such situations, the operator selects the next behavior for the robot. After this operation, developers updated the word dictionaries for speech recognition and the episode rules based on the operation histories so that the robot can autonomously select its next behavior in the future.

2.4.8.3 Knowledge Provider

With the current technology, only humans can provide knowledge to the robot. Developers input knowledge in advance as a form of behavior. But this in-advance effort is limited to what the developers can expect; in reality, much unexpected knowledge is needed. For example, although the robot has behaviors for all restaurants, when asked about a Japanese-food restaurant, the robot couldn't say something like, "There are two Japanese-food restaurants: a *sushi* restaurant and a *soba* restaurant. Which do you prefer?" For such a case, the operator directly typed the sentence so that the robot could respond. Later, developers added the appropriate behaviors for the situation.

2.4.9 Conversational Fillers

The operator roles include decision making so the operator needs a few seconds to manage the robot if the question is complex or difficult. However, since users might feel uncomfortable during slow responses or long pauses, robot response time is critical. To solve such problems, we implemented a *conversational filler* to buy time [29]. When the operator needs a few seconds, he/she executes a *conversational filler* behavior to notify listeners that the robot is going to respond soon.

2.4.10 Field Trial

2.4.10.1 Procedure

A field trial was conducted at a large, recently built shopping mall consisting of three floors for shopping and one for parking with approximately 150 stores and a large supermarket. The robot was placed in a main corridor of the mall weekday afternoons from 1 to 5 for 5 weeks (from July 23 to August 31, 2007, except for a busy week in the middle of August). This schedule was decided based on an agreement with the mall management to avoid busy times to prevent situations where too many people might crowd around the robot.

The robot was open to all visitors. Those who signed up for the field trial (participants) received a passive-type RFID embedded in a cell phone strap (Figure 2.24). We recruited these participants by two methods: (1) a flyer distributed to residents around the mall, and (2) on-site sign up during the first 3 weeks while our staff approached visitors who seemed interested in the robot. The participants filled out consent forms when they enrolled and questionnaires after the field trial. They were not paid, but they were allowed to keep their RFID tags.*

2.4.11 Results

2.4.11.1 Overall Transition of Interactions

Figure 2.27 shows the number of interactions the robot engaged in. Here, one interaction represents an interaction that continued with the visitor until the robot said goodbye. Figure 2.28 shows the interaction scenes (supplement video file contains other scenes of interaction). During the first 3 weeks our staff invited visitors for registration and interaction with the robot. From the fourth week onward, our staff stood near the robot for safety. There was

---

* The experimental protocol was reviewed and approved by our institutional review board.

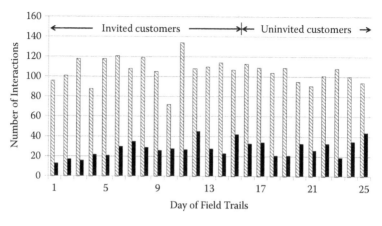

FIGURE 2.27
Daily visitors and participants.

an average of 105.7 interactions each day. As the graph shows, the number of interacting persons did not differ over the 5-week period. Multiple persons interacted with the robot (an average of 1.9 persons per interaction).

There were 332 participants who signed up for the field trial and received RFID tags; 37 participants did not interact with the robot at all, 170 participants visited once, 75 participants visited twice, 38 visited three times, and 26 visited four times. The remaining 23 participants visited from five to 18 times. On average, each participant interacted 2.1 times with the robot, indicating that they did not repeat interaction very much. One obvious shortage was the trial duration; since every day many nonparticipant visitors waited in line to interact with the robot, some participants reported that they hesitated to interact with the robot because it was too crowded. Figure 2.27 shows the number of participants who interacted each day, with an average of 28.0 persons per day.

2.4.11.2 Perception of Participants

When the field trial finished, we mailed questionnaires to 332 participants and received 235 answers. All items were on a 1 to 7 point scale where 7 represents the most positive, 4 represents neutral, and 1 represents the most negative.

2.4.11.2.1 Impression of Robot

The questionnaire included items such as "Intention of use" (studied in [30]), "(the degree of) Interest," "Familiarity," and "Intelligence," which resulted in respective scores of 5.0, 4.9, 4.9, and 5.1 (S.Ds. were 1.3, 1.4, 1.4, and 1.4). Many positive, free-answer form comments described the robot as cute and friendly.

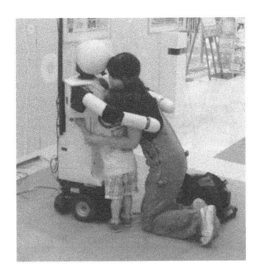

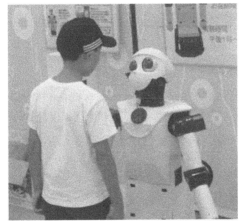

FIGURE 2.28
Interaction scenes.

2.4.11.2.2 Route Guidance

The questionnaire answers about the adequacy of route guidance resulted in an average of 5.3 points (S.D. was 1.3). In a free-description form, the following comments were made:

> "The robot correctly answered when I asked about a particular shop."
>
> "I'm surprised that its route guidance was so detailed."
>
> "Its route guidance was appropriate and very easy to understand."
>
> "The robot was useful for questions that I hesitated to ask because they seemed too simple."

2.4.11.2.3 Providing Information

The questionnaire answers about the usefulness and interest in the information resulted in an average of 4.6 and 4.7 points (S.D.s were 1.4 and 1.3). Moreover, 99 out of 235 participants reported that they visited a shop mentioned by the robot, and 63 participants bought something based on the information provided by the robot. We particularly asked about reasons in a free-description form and received the following comments:

> "The robot recommended a kind of ice cream that I hadn't eaten before, so I wanted to try it."
>
> "The movie mentioned by the robot sounded interesting."
>
> "Since Robovie repeatedly mentioned crepes, my child wanted to eat one."

These results suggest that the robot's information-providing function affected them, increased their interest in particular shops and products, and even encouraged them to actually buy products.

2.4.11.2.4 Building Rapport

The questionnaire answers about degree of perceived familiarization resulted in a 4.6 point on average (S.D. was 1.5). In the free-description form, comments included:

> "Since it said my name, I felt the robot was very friendly."
>
> "The robot was good since it seemed as if it gradually got familiar with me."
>
> "I'm surprised that the robot has such a good memory." (People in the United States also perceive that robots are good at memorization [31].)
>
> "My child often said let us go to the robot's place, and this made visiting the mall more fun."
>
> "The robot was very friendly. I went with my 5-year-old daughter to interact with the robot; on the last day, she almost cried because it was so sad to say goodbye. She remembers it as an enjoyable event: at home, she imitates the robot's behavior and draws pictures of it."

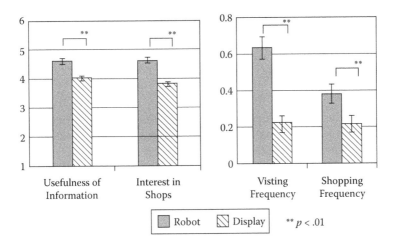

FIGURE 2.29
Comparison of robot and a display.

2.4.11.3 Comparison with an Information Display

We asked participants how often they were influenced by information displays in the same mall (Figure 2.22). In the questionnaires, participants were asked to answer the following: "Usefulness of information provided by display/robot," "Interest in shops mentioned by display/robot," "Visiting frequency triggered by display/robot," and "Shopping frequency triggered by display/robot." The order of the questionnaires about the display and robot was counterbalanced.

Figure 2.29 shows the comparison result. There were significant differences (F(1,229) = 40.96, 69.52, 36, 19, and 7.66, $p < .01$ for all four items). Thus, for the participants, the robot provided more useful information and elicited more shopping.

Note that this is not like laboratory experiments with precise control. Thus, the comparisons have the following unbalanced factors. Yet we believe the comparisons are still useful to understand the phenomena caused by the robot.

Duration of comparison: Regarding the robot, we asked about their experience during the field trial. In regard to the display, to include its possible novelty effect (it could be novel to them, as it is very large), we asked them to answer their experience regarding the 4 months of the duration (from the opening of the mall until the end of the field trial).

Way of providing information: The display shows information about a shop by highlighting information. The target shop is switched about once a minute. Note that since this display is in a commercial-based service, we assume that it is well prepared.

Participant interest: The participants might be more interested in the robot than the other mall visitors, since we suspect that participation in the field trial reflected interest in the robot. However, this is a limitation of our study as a field trial, which needed spontaneous participation; for example, they would have had to register for the RFID tags.

2.4.11.4 Integrated Analysis

2.4.11.4.1 Structural Equation Modeling (SEM)

We analyzed the relationships among impression, perceived usefulness, and the affect on shopping behavior using structural equation modeling (SEM), which is a relatively new statistical analysis for revealing the relationships behind observed data. Its process resembles factor analysis to reveal latent variables and regression analysis to associate variables to produce a graphical model of causal-result relation. Since SEM is an established technique with many textbooks such as [32,33], we leave further explanation of this technique to these textbooks. The following paragraphs report how we applied this technique to our data.

For the modeling, our hypothesis is that their interaction experiences with the robot (observed as impression and day of visit) affected their shopping behavior as an advertisement effect. We made a model that included the latent variables of advertisement and interest effects as possible consequences. We added the latent variables of the impressions of the robot and established rapport (relationships) with it as well as the experience of shopping as possible causal factors.

2.4.11.4.2 Analysis Result

Figure 2.30 shows the best-fit model produced by SEM. In the figure, for readability we didn't draw error variables that are only associated with one variable. The variables in the squares are the observed variables (such as the questionnaire items), and those in the circles are the latent variables retrieved by the analysis (named by us). The numbers around the arrows (path) are the values of the path coefficients, similar to coefficients in regression analysis. The numbers on the variables show the coefficient of determination, R^2. Thus, 30% of the "Advertisement effect" is explained by the factors of "Relationships with robot" and "Experience of shopping," and 42% of the "Interest effect" is explained by the factor of "Impression of robot."

Regarding the model's validity, this analysis result shows good fitness in the appropriateness indicators of GFI = .957, AGFI = .931, CFI = .987, and RMSEA = .028. (According to [32], the desired range of the indicators should be as follows: GFI, AGFI ≥ .90, CFI ≥ .95, and RMSEA ≤ .05). Each path coefficient is significant at a significance level of 1%.

In SEM analysis, there is an indicator, AIC, for the best-fitness of this model. The model with the minimum AIC value is considered the best among the models with the same variables. The analysis result of Figure 2.30 has a minimum AIC value of 115.9. For example, a model with one extra path from "Impression of robot" to "Advertisement effect" results in an AIC value of 116.9,

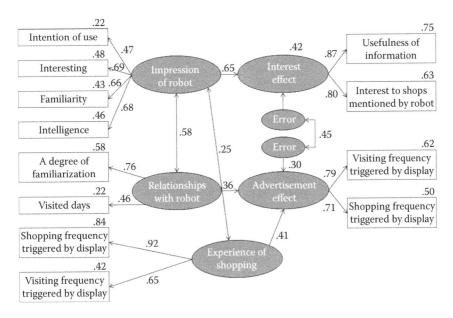

FIGURE 2.30
Retrieved model about observed variables.

so this path itself is not significant (coefficient = –.10, *p* = .36). This suggests that "Advertisement effect" is not directly affected by "Impression of robot."

2.4.11.4.3 Interpretation

The interpretation of this modeling result is quite interesting. The model suggests that the participants who positively evaluated the impression of the robot tended to be positive about the interest effect (coefficient = .65); however, the advertisement effect is not associated with the impression of the robot, but with the relationships with it (coefficient = .36). Thus, the factor of the relationships with the robot explains 13% of the deviation of the advertisement effect. Although this ratio might not be so high, we believe that it is interestingly high for such shopping behavior, since shopping behavior largely depends on people's various situations (financial, interests, time, occasion, etc.). It implies that development of relationships with the robot would increase the advertisement effect. Although to increase the relationships, impression could be important.

2.4.12 Discussion

2.4.12.1 Degree of Operator Involvement

Since this study was conducted with operators, it is useful to show how often the robot was under their control. Figure 2.31 shows the number of operations. As described in Section 2.4.8, one operator role was to "Substitute

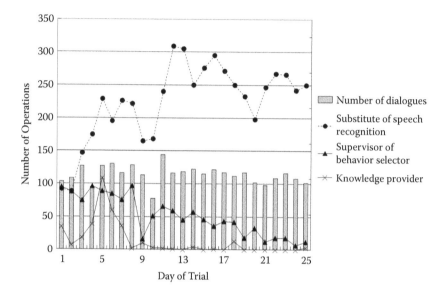

FIGURE 2.31
Number of operations by operator.

speech recognition," which we expect to be automated in the future. The operator did this two or three times per dialogue.

In contrast, the result shows that the operator's load for the remaining two roles, "Supervisor of behavior selector" and "Knowledge provider," gradually decreased. This result is promising, because these two roles will be difficult to do autonomously. After day 10, 254.2 "Substitute of speech recognition," 1.7 "Knowledge provider," and 13.4 "Supervisor of behavior selector" operations were conducted per day.

During the field trial, we continued to implement the interactive behaviors to supplement the missing knowledge that the operators needed to operate. On average, we added 0.2 interactive behaviors to reduce the "Knowledge provider" task and 3.4 rules for transition among behaviors to reduce the "Supervisor of behavior selector" task per day. This result shows one promising case of robot development that operates in a real field under the supervision of human operators.

2.4.13 Conclusion

We developed a robot that was designed to provide information such as route guidance and other shopping information for a shopping mall. A 5-week field trial was conducted in a shopping mall. We recruited and registered participants for RFID tags and gave them questionnaires after the field trial. Analysis results indicate that they accepted the robot with positive impressions and were influenced by the information provided by it. The comparison shows that the robot more successfully invited visitors for shopping than an information-presenting display. Integrated analysis revealed the importance of establishing relationships between customers and the robot for larger advertisement effects on shopping behavior. The robot performed well in the information-providing task using gestures and natural language and successfully influenced people's daily shopping activities. In contrast, the study failed to show whether the robot could elicit spontaneous repeated interactions; a limited number of participants visited repeatedly. This aspect should be explored more in future studies.

Acknowledgments

We wish to thank the administrative stuff at the Takanohara AEON for their cooperation. We wish to thank Dr. Akimoto, Dr. Miyashita, Dr. Sakamoto, Mr. Glas, Mr. Tajika, Mr. Nohara, Mr. Izawa, and Mr. Yoshii for their help. This research was supported by the Ministry of Internal Affairs and Communications of Japan.

References

1. Kidd, C. and Breazeal, C., Effect of a robot on user perceptions, *IROS'04*, pp. 3559–3564, 2004.
2. Powers, A., et al., Comparing a computer agent with a humanoid robot, *HRI2007*, pp. 145–152, 2007.
3. Mutlu, B., et al., A Storytelling robot: Modeling and evaluation of human-like gaze behavior, *IEEE Int. Conf. on Humanoid Robots (Humanoids2006)*, pp. 518–523, 2006.
4. Kahn, P. H., Jr., et al., Design patterns for sociality in human robot interaction, *HRI2008*, pp. 97–104, 2008.
5. Burgard, W., et al., The interactive museum tour-guide robot, *Proc. of National Conference on Artificial Intelligence*, pp. 11–18, 1998.
6. Siegwart, R., et al., Robox at Expo.02: A large scale installation of personal robots, *Robotics and Autonomous Systems*, 42(3), pp. 203–222, 2003.
7. Kanda, T., et al., Interactive robots as social partners and peer tutors for children: A field trial, *Human Computer Interaction*, 19(1–2), pp. 61–84, 2004.
8. Wada, K., et al., Effects of robot-assisted activity for elderly people and nurses at a day service center, *Proceedings of the IEEE*, 92(11), pp. 1780–1788, 2004.
9. Dautenhahn, K., et al., A quantitative technique for analysing robot-human interactions, *IROS'02*, pp. 1132–1138, 2002.
10. Kozima, H., Nakagawa C., and Yasuda, Y., Interactive robots for communication-care: A case-study in autism therapy, *Ro-Man 2005*, pp. 341–346, 2005.

11. Tanaka, F. et al., Socialization between toddlers and robots at an early childhood education center, *Proc. of the National Academy of Sciences of the USA*, 104(46), pp. 17954–17958, 2007.
12. Dahlback, D., et al., Wizard of Oz studies—why and how, *Knowledge based systems*, 6(4), pp. 258–266, 1993.
13. Green A., et al., Applying the wizard-of-Oz framework to cooperative service discovery and configuration, *Ro-Man 2004*, 2004.
14. Glas, D. F., et al., Simultaneous teleoperation of multiple social robots, *HRI2008*, pp. 311–318, 2008.
15. Dario, P., Guglielmelli, E., and Laschi, C., Humanoids and personal robots: Design and experiments, *Journal of Robotic Systems*, 18(12), pp. 673–690, 2001.
16. Goetz, J., Kiesler, S., and Powers, A., Matching robot appearance and behaviors to tasks to improve human robot cooperation, *Ro-Man 2003*, pp. 55–60, 2003.
17. Robins, B., Ferrari, E., and Dautenhahn, K., Developing scenarios for robot assisted play, *Ro-Man2008*, pp. 180–186, 2008.
18. Research study for the scope of the strategy map of robotics technology, 2005. (Available at http://www.nedo.go.jp/database/index.html, with index code 100007875) (in Japanese).
19. Picard, R. W., *Affective Computing*, The MIT Press, Cambridge, MA, 1997.
20. Kanda, T., et al., Who will be the customer? A social robot that anticipates people's behavior from their trajectories, *UbiComp2008*, 2008.
21. Sanfeliu, A., Hagita, N., and Saffiotti, A., Special issue: Network robot systems, *Robotics and Autonomous Systems*, 2008.
22. Ishi, C. T., et al., Robust speech recognition system for communication robots in real environments, *IEEE Int. Conf. on Humanoid Robots*, pp. 340–345, 2006.
23. Shiomi, M., et al., A semi-autonomous communication robot—A field trial at a train station—, *HRI2008*, pp. 303–310, 2008.
24. Bickmore, T. W., and Picard, R. W., Establishing and maintaining long-term human-computer relationships, *ACM Transactions on Computer-Human Interaction*, Vol. 12, No. 2, pp. 293–327, 2005.
25. Sugiyama, O., et al., Humanlike conversation with gestures and verbal cues based on a three-layer attention-drawing model, *Connection Science*, 18(4), pp. 379–402, 2006.
26. Gockley, R., Forlizzi, J., and Simmons, R., Interactions with a moody robot, *HRI2006*, pp. 186–193, 2006.
27. Murakita, T., et al., Human tracking using floor sensors based on the Markov chain Monte Carlo Method, *Proc. Int. Conf. Pattern Recognition (ICPR04)*, pp. 917–920, 2004.
28. Kanda, T., et al., A two-month field trial in an elementary school for long-term human-robot interaction, *IEEE Transactions on Robotics*, 23(5), pp. 962–971, 2007.
29. Shiwa, T., Kanda, T., Imai, M., Ishiguro, H., and Hagita, N., How quickly should communication robots respond? *HRI2008*, 2008.
30. Heerink, M., Kröse, B., Wielinga, B., and Evers, V., Enjoyment intention to use and actual use of a conversational robot by elderly people, *HRI2008*, pp. 113–119, 2008.
31. Takayama, L., Ju, W., and Nass, C., Beyond dirty, dangerous and dull: What everyday people think robots should do, *HRI 2008*, pp. 25–32, 2008.
32. Toyoda, H., Structural equation modeling, Tokyo Tosho, 2007 (in Japanese).
33. Kaplan, D. W., *Structural Equation Modeling: Foundations and Extensions*, Sage Publications, Newbury Park, CA, 2000.

3

Users' Attitude and Expectations

3.1 Introduction

The studies in Chapter 2 reported how users interacted with social robots. Some users showed very positive attitudes trying to interact with robots repeatedly; some users showed an intermediate attitude, hesitantly crowding around the robot and watching the robot's behavior, but did not directly interact with it. Why would such difference in attitude occur?

In human–robot interaction, a number of research works have analyzed users' expectations. People have expectations about robots' jobs and tasks, and how robots should be designed. It was found that people prefer human-like appearance for jobs that require social skills [1]. It is suggested that people would delegate tasks to human-like robots rather than to machine-like robots [2]. Compared with humans, people expect robots to work for service-oriented tasks or tasks that need keen perception [3], and troublesome tasks [4]. Cultural influence was also found, as in some cultures people have more of an acceptance attitude toward robots than found in other culture [5]. Further, cultural differences would create a preferred way for a robot to interact with people [6].

In this chapter, we introduce our work related to users' attitude and expectations toward robots. The first work addresses phenomena in which a robot is known to be teleoperated. While we consider the needs of teleoperation in the deployment phase (Chapter 6 discusses motivation and technical issues more), it is not clear the impact on users. Some people would argue that people would only interact with autonomous robots but not with teleoperated robots. Is this a valid concern? The work in Chapter 3.2 will address this question.

The second study addresses people's attitude. While a number of studies addressed the design of robots that would influence people's acceptance attitude, not many studies addressed the potential reason why some people would show negative attitude toward robots. The study deals with such a problem. It reports psychological scales to measure people's attitude, and provides insight into why people would have negative attitude toward robots.

With these two studies, we depict a diversity of issues related to people's expectations and attitudes. First, there is a study that addresses concern about the potential relationship between people's expectation and their attitudes.

Second, there is a study that focuses on only attitude, revealing relationships between people's internal state and behaviorally expressed attitude. We included these two studies as they would fit with the context of the book well, that is, robots that interact with people in a real world supported with a networked system. Moreover, these studies focused on the immediate consequences of technical construction reported in our study. However, it is not our intention to exclude other issues such as design related to appearance, which are also very important in making robots socially acceptable with settings appropriate expectations.

Reference

1. Goetz, J., Kiesler, S., and Powers, A. (2003). *Matching Robot Appearance and Behavior to Tasks to Improve Human-Robot Cooperation*. Paper presented at the IEEE Int. Workshop on Robot and Human Interactive Communication (RO-MAN2003).
2. Hinds, P. J., Roberts, T. L., and Jones, H. (2004). Whose job is it anyway? A study of human-robot interaction in a collaborative task. *Human-Computer Interaction, 19*(1), 151–181.
3. Takayama, L., Ju, W., and Nass, C. (2008). *Beyond Dirty, Dangerous and Dull: What Everyday People Think Robots Should Do*. Paper presented at the ACM/IEEE Int. Conf. on Human-Robot Interaction (HRI2008).
4. Hayashi, K., Shiomi, M., Kanda, T., and Hagita, N. (2010). *Who Is Appropriate? A Robot, Human and Mascot Perform Three Troublesome Tasks*, IEEE Trans. on automated mental developments, 4(2), 150–160.
5. Bartneck, C., Suzuki, T., Kanda, T., and Nomura, T. (2006). The influence of people's culture and prior experiences with Aibo on their attitude towards robots. *AI & Society, 21*(1–2), 217–230.
6. Wang, L., Rau, P.-L. P., Evers, V., Robinson, B. K., and Hinds, P. (2010). *When in Rome: The Role of Culture & Context in Adherence to Robot Recommendations*. Paper presented at the ACM/IEEE Int. Conf. on Human-Robot Interaction (HRI2010).

3.2 Is Interaction with Teleoperated Robots Less Enjoyable?

Fumitaka Yamaoka,[1,2] *Takayuki Kanda,*[1]
Hiroshi Ishiguro,[1,2] *and Norihiro Hagita*[1]

ABSTRACT

Robots in networked robot systems are sometimes teleoperated by people. Our question is whether the person interacting with such teleoperated robots feels that he/she is interacting with the robot itself or a person behind it. How do such beliefs affect interaction? We conducted an experiment to study the effect of a human presence behind a robot. We used an interactive robot which participants can bodily interact with. They were instructed before the interaction that they would either be interacting with a robot that is autonomous controlled by a program or teleoperated by a person; in reality, in both conditions the robot was autonomous. Experimental results indicated that two-thirds of the participants felt that they were interacting with the robot itself regardless of the instruction. Their enjoyment was unaffected by the knowledge that the robot was controlled by a program or a person, although their impression of robot intelligence indicated that they distinguished these conditions. On the contrary, the remaining one-third of the participants, who reported that they were interacting with a person behind the robot, were affected by the knowledge; people who were told they were interacting with an autonomous robot perceived more enjoyment than people who were told they were interacting with a teleoperated robot.

3.2.1 Introduction*

The robot used in the field trial in Chapter 2.4 was teleoperated. There are some other examples of teleoperated social robots [1,2,3]. Some operated robots are being developed as a prototype of autonomous robots, known as the WOZ (Wizard of OZ) approach [4,5,6].

However, there seems to be a long-standing belief that people expect to interact with autonomous robots, but not a robot that is teleoperated. One could argue that people interaction would be significantly different between interaction with autonomous robots and teleoperated robots. One could even argue that people would be disappointed if they know a robot being teleoperated. To what extent is such a belief true?

Several researchers have reported how people are affected by the presence of humans in human–computer interaction. For instance, Reeves and Nass reported that people are usually unaware of humans behind computers; however, when a malfunction happens, they become aware of programmers since they want to complain [7]. In contrast, Yamamoto et al. reported that people perceived less enjoyment and spent less time with a teleoperated system than with an autonomous system, which performs a simple word game called *shiritori* [8]. Will this be replicated in human–robot interaction?

3.2.2 Experimental System

We used our previously developed interactive humanoid robot system [9]. The robot was developed for bodily nonverbal interaction, based on the

---

* This chapter is a version of a previously published paper, modified to be comprehensive and fit with the context of this book, by Yamaoka, F., Kanda, T., Ishiguro, H., & Hagita, N. (2007), *Interacting with a Human or a Humanoid Robot?* Paper presented at the IEEE/RSJ Int. Conf. on Intelligent Robots and Systems (IROS2007).

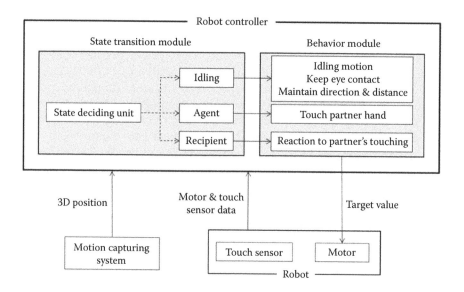

FIGURE 3.1
System configuration. (a) Idling motion, (b) keeping eye contact/maintaining distance and direction, (c) touching partner's hand, (d) reacting to partner's touching.

findings in developmental psychology [10] reporting on what is considered as animate. The robot reacts to a person's approaching motion before being touched (recipient state), and initiates an approach to him/her if it is not approached (agent state).

The system consists of a humanoid robot, Robovie, a motion-capturing system, and a robot controller software (Figure 3.1). A motion-capturing system tracks the position of markers attached to the head, arms, and body of both the robot and the interacting person, which are processed by the robot controller.

In the robot controller, the state controller module chooses the robot's internal state. The robot's internal state is chosen from three states, *recipient, idling,* and *agent*, according to the partner's behavior (Figure 3.2). When its partner tries to touch the robot, it chooses *recipient* state and react to touching by executing one of "reaction to partner's touch" behavior. Otherwise it starts *idling* state to wait for a partner's touch. In the *idling* state, it performs *idling motion* while *maintaining eye contact,* and *maintaining distance and direction.* If the *idling* state continues more than five seconds, it transits to *agent state* and performs "touching the partner's hand" for five seconds. If the partner reacts, the internal state changes from *agent* to *recipient.* Further details are described in [9].

3.2.3 Experiment

We followed the experimental framework of the word game experiment [8]. In our experiment, participants were instructed that they would be interacting

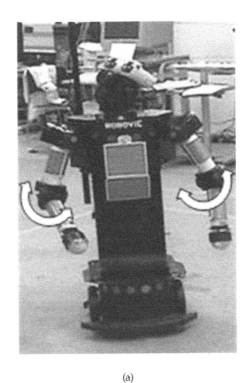

(a)

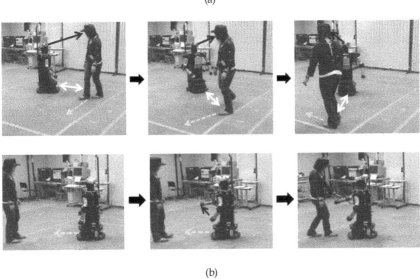

(b)

FIGURE 3.2
Behaviors: (a) Instructions in autonomous condition. (b) Instructions in operated condition.

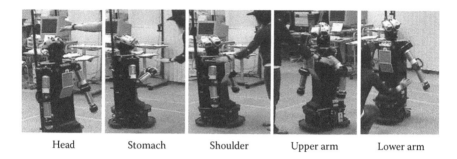

| Head | Stomach | Shoulder | Upper arm | Lower arm |

FIGURE 3.2 (continued)

with a robot controlled by either a program or a person. Then, the participant interacted with the robot for a while. Interaction was interrupted to fill out a questionnaire. While passing the time until the next experiment, the participant was offered the opportunity to either interact more with the robot or just wait. With this procedure, we measured voluntary interaction time as an objective measure of enjoyment.

3.2.3.1 Participants and Environment

There are 77 university students participating in this experiment (50 men, 27 women). They were paid for the participation. The experiment was conducted in a 7.5 m × 10.0 m room. Due to the limitations of the motion-capturing system, participants only interacted with the robot within an area of 2 m × 2 m.

3.2.3.2 Procedure

We performed the experiment in the five steps below. Waiting in Step 3 is preparation for measuring the extension time that shows participants' implicit feelings of enjoyment, that is, if humans enjoy interacting with the robot, they should continue to play longer.

Step 1: Instructions before Experiment

Participants were informed about the control method of the robot and how to interact with the robot by watching an explanation video (Figure 3.3). The type of control method they were presented reflected the condition to which they were assigned. Next, we instructed them to interact with the robot, such as approaching it, or stepping away from it, touching its body, or observing its motions and reactions.

Step 2: First Interaction and Evaluation

Participants interacted with the robot for three minutes (Figure 3.4). After the interaction, participants filled questionnaires to provide their impression.

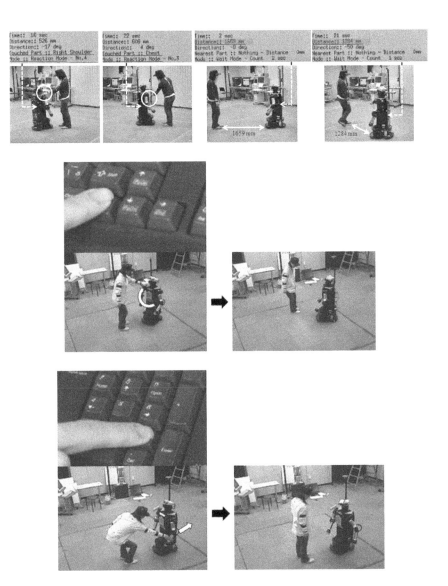

FIGURE 3.3
Instruction video.

Step 3: Offering Opportunity to Have Additional Interaction

The experimenter told participants that they must wait for about 50 minutes for the next experiment due to preparation delays. The experimenter also invited them to play with the robot while waiting. Participants who chose to play with the robot were instructed to stop playing when they felt bored. If they interacted with the robot longer than 20 minutes, the experimenter stopped them.

FIGURE 3.4
Scene of the experiment.

Step 4: Debriefing

Participants were informed about the idea behind the experiment and the fact that the robot was operated autonomously during the entire experiment.

3.2.3.3 Conditions

Participants interacted with the robot in one of the following two conditions:

Autonomous: Participants were informed that the robot moved autonomously. The explanation video in the autonomous condition showed how the information concerning partner (distance, direction, and touched part, etc.) is processed in the program and how the robot's internal state changes its behavior to reflect the information (Figure 3.3a).

Operated: Participants were informed that the robot was operated by an operator. The explanation video in the *operated* condition deceptively informed that the robot was operated with keyboard entries by a human operator (Figure 3.3b).

In both conditions, the robot was operated autonomously with the program reported in the Section 3.2.

3.2.3.4 Measures

We evaluated the robot based on the following participants' impressions and behavior.

Impressions of the robot: Participants rated impressions of *autonomy, intelligence*, and *enjoyment on* a 1-to-7 point Likert scale, where 1 is the lowest and 7 is the highest. This is measured at step 2 in the experimental procedure.

Time of additional interaction: We measured time of additional interaction, measured as the time they spend during offered additional interaction at step 3 in the experimental procedure. We consider that this reflects the degree of enjoyment participants perceived.

Attribution of interaction: After step 3, participants filled a questionnaire whether they attributed interaction to the humanoid robot itself or a person behind the robot on a 1-to-7 scale. When participants most felt they were interacting with the robot itself, they gave 1 point; they gave 7 points when they most felt they were interacting with a person behind the robot.

3.2.4 Results

3.2.4.1 Does Prior Knowledge of Operator's Presence Vary Impressions?

Figure 3.5 shows the results of impressions and interaction time. We conducted a one-way analysis of variance (ANOVA). It reveals significant difference only in *intelligence* ($F(1,75) = 9.674$, $p = .003$, $\eta^2 = .114$), and no significant difference found in *autonomy* ($F(1,75) = 2.611$, $p = .110$, $\eta^2 = .034$). This result shows that participants attributed the differences of experimental conditions to the robot's intelligence, but not to autonomy. Participants seemed to perceive that the robot was less intelligent if it were being operated. This at least indicates that the experimental control was successful.

The result show no significant difference across condition in *enjoyment* ($F(1,75) = 0.002$, $p = 0.962$, $= \eta^2 = 000$) and *time of additional interaction* ($F(1,75) = .970$, $p = .328$, $\eta^2 = .013$). As indicated, effect size is very small (while effect size of .100 is considered as "small" effect, $\eta^2 = .000$ in *enjoyment* and .013 in *time of additional interaction*), even though there could be a difference across condition. This suggest that, on average, the prior knowledge on the presence of an operator did not affect enjoyment in interacting with a robot.

There were also no significant difference in *attribution of interaction* ($F(1,75) = 1.990$, $p = .162$, $\eta^2 = .026$). Thus, the way people subjectively perceived human presence behind the robot was not affected (or only affected very little) by knowing whether they are informed the operator's presence.

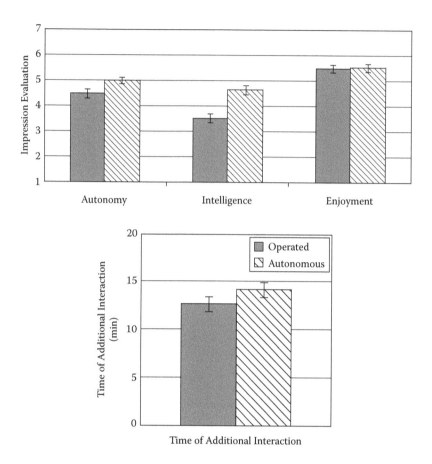

FIGURE 3.5
Result of impression evaluation and behavior with analysis.

3.2.4.2 Participants Affected by Prior Knowledge of Operator's Presence

Prior knowledge did not influence enjoyment on average. However, it is often argued that people would enjoy it less if a robot is teleoperated. We assume that a minority of people would perceive differently. After examining the data, we found that participants' *attribution of interaction* is an indicator to identify such people whose enjoyment was harmed from prior knowledge of an operator's presence.

We classified participants into the two groups. The first group, the *robot attribution* group, is the largest (29 participants in *operated*, 25 in *autonomous* condition) and perceived that they were interacting with the robot. They gave *attribution of interaction* scores less than four. The second group, the *human attribution* group, is the minority (13 participants in *operated*, 10 in *autonomous* condition) who perceived that they were interacting with a person behind the robot.

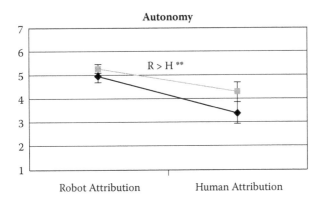

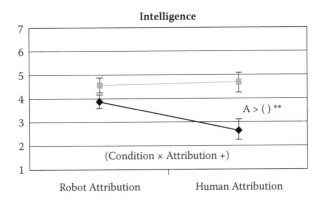

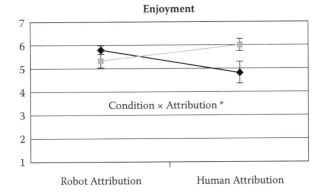

FIGURE 3.6
The result of behavior with the analysis.

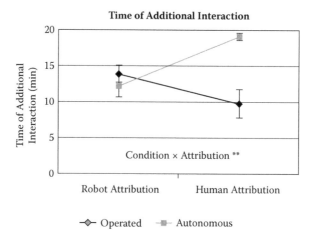

FIGURE 3.6 (continued)

Figure 3.6 shows the results of participant impressions and interaction time. A two-way (2 prior knowledge × 2 attributions) ANOVA revealed interaction effect across two factors summarized below:

- *Autonomy*—Only the *attribution* factor was significant (F(1,73) = 15.714, $p < .001$, $\eta^2 = .177$). This rather intuitive result indicates that people who attributed interaction partner as a human evaluated the robot less autonomously. There is some significance in the *prior knowledge* factor (F(1,73) = 3.624, $p = .061$, $\eta^2 = .047$).

- *Intelligence*—There was a significant difference in the *prior knowledge* factor (F(1,73) = 12.737, $p = .001$, $\eta^2 = .149$). There was almost significant interaction: (F(1,73) = 2.948, $p = .090$, $\eta^2 = .039$). Analysis of interaction revealed a significance difference between *operated* and *autonomous* conditions in *human* attribution ($p = .002$).

- *Enjoyment*—There was significant interaction in *enjoyment* (F(1,73) = 6.748, $p = .011$, $\eta^2 = .085$, at the time of additional interaction). Analysis of interaction revealed a significance difference between *operated* and *autonomous* conditions in *human* attribution ($p = .031$), and a significance difference between *human* and *robot* attribution in *operated* condition ($p = .026$).

- *Time of additional interaction*—There was a significant interaction (F[1,73] = 11.767, $p = .001$, $\eta^2 = .139$) and *prior knowledge conditions* (F[1,73] = 5.600, $p = .021$, $\eta^2 = .071$). Analysis of interaction revealed a significant difference between *operated* and *autonomous* conditions in *human* attribution ($p = .001$), and a significant difference between *human* and *robot* attribution in *autonomous* condition ($p = .005$) and an almost significant difference in *operated* condition ($p = .059$).

Overall, results indicated that people in *robot attribution* group were mostly unaffected by the *prior knowledge of the operator's presence*. In contrast, people in *human attribution* group were affected much more. If they are informed the presence of human operator behind the robot, they perceived less enjoyment and spent less time. Interestingly, such people spent more time if they were informed that the robot is autonomous.

3.2.5 Conclusion

We conducted this study aiming to reveal to what extent teleoperation would affect people's interaction with a social robot. In a bodily nonverbal interaction context, we found that, on average, people's enjoyment was unaffected, though they perceived the teleoperated robot as less intelligent. Two-thirds of the participants perceived that they were interacting with the robot itself, an impression that contributed to the above-average impression. In contrast, the remaining minority, that is one-third of participants who perceived that they were interacting with a human behind the robot, was affected by prior knowledge of the presence of an operator. They are the people who enjoyed spending more time in interaction when they were informed that the robot is autonomous, and spent less time and enjoyed the robot less when they were informed that the robot was operated by a person.

We interpret the result that teleoperation would overall not harm people's enjoyment in interacting with the robot. If a robot were used for an intelligent task, e.g., providing some complex information, perhaps the decrease of intelligence would be affected, though it is a limitation of the study that we only tested bodily nonverbal interaction; thus, we would need to conduct further study on how prior knowledge of operator's presence would influence verbal interaction and task performance.

Acknowledgments

This research was supported by the Ministry of Internal Affairs and Communications of Japan.

References

1. Iwamura, Y., Shiomi, M., Kanda, T., Ishiguro, H., and Hagita, N. (2011). Do elderly people prefer a conversational humanoid as a shopping assistant partner in supermarkets? Paper presented at the *6th ACM/IEEE International Conference on Human-Robot Interaction (HRI2011)*.

2. Shiomi, M., Kanda, T., Glas, D. F., Satake, S., Ishiguro, H., and Hagita, N. (2009). Field trial of networked social robots in a shopping mall. Paper presented at the *IEEE/RSJ Int. Conf. on Intelligent Robots and Systems (IROS2009)*.
3. Tanaka, F., Cicourel, A., and Movellan, J. R. (2007). Socialization between toddlers and robots at an early childhood education center. Paper presented at the *Proceedings of the National Academy of Sciences of the USA (PNAS)*.
4. Green, A., Hüttenrauch, H., and Severinson Eklundh, K. (2004). Applying the Wizard-of-Oz framework to cooperative service discovery and configuration, in *Proceedings of the 13th IEEE International Workshop on Robot and Human Interactive Communication (Ro-Man 2004)*, pp. 575–580.
5. Walters, M. L., Dautenhahn, K., Boekhorst, R., Koay, K. L., Kaouri, C., Woods, S., Nehaniv, C., Lee, D., and Werry, I. (2005). The influence of participants' personality traits on personal spatial zones in a human-robot interaction experiment, *IEEE International Workshop on Robot and Human Communication (Ro-Man 2005)*, pp. 347–352.
6. Shiomi, M., Kanda, T., Koizumi, S., Ishiguro, H., and Hagita, N. Group attention control for communication robots with Wizard of Oz approach. Paper presented at the ACM/IEEE Int. Conf. on Human Robot Interaction. (HRI 2007).
7. Reeves, B. and Nass, C. (1996). *The Media Equation*. How People Treat Computers, Television, and New Media like Real People and Places. Cambridge Univ. Press, New York, NY.
8. Yamamoto, Y., Matsui, T., Hiraki, K., Umeda, S., and Anzai, Y. (1994). Interaction with a compute system, *Journal of Japanese Cognitive Science Society*, 1(1), 107–120 (in Japanese).
9. Yamaoka, F., Kanda, T., Ishiguro, H., and Hagita, N. (2005). Lifelike behavior of communication robots based on developmental psychology findings, *IEEE International Conference on Humanoid Robots (Humanoids2005)*, pp. 406–411.
10. Rakison, D. H., and Poulin-Dubois, D. (2001). Developmental origin of the animate-inanimate distinction. *Psychological Bulletin*, 127(2), 209–228.

3.3 Hesitancy in Interacting with Robots— Anxiety and Negative Attitudes

Tatsuya Nomura, Takayuki Kanda, Tomohiro Suzuki, and Kensuke Kato

ABSTRACT

We aim to identify people's psychology behind their behavior toward social robots. First, two psychological scales were developed, each corresponding with people's psychological state related to *anxiety* and *attitude*: the Negative Attitudes toward Robots Scale (NARS) and the Robot Anxiety Scale (RAS). While *attitude* represents rather sustainable predisposition, *anxiety* involves a short-lasting *state* and person-dependent *trait*. We conducted an empirical study to reveal the relationships between people's attitudes and anxiety, and their behavior toward a robot, particularly communication avoidance behavior. In the experiment, participants and a humanoid robot are engaged in simple interactions, including scenes of meeting, greeting, self-disclosure, and physical contact. Experimental results confirmed the relationship between negative attitudes and anxiety, and communication avoidance behavior.

3.3.1 Introduction*

While we have observed a number of people elicited to interact with social robots in real fields, there are a certain number of people who expressed their hesitation to interact with a social robot. They are seemingly not busy. Some of them seem to like the robot, curiously watching robots' interaction with other people. Yet, when a robot approached to them, they typically expressed hesitation and declined interacting with the robot. What made such difference in their attitude? Why do some people show hesitation while some do not?

Hesitation would be not a problem when a robot is used for entertainment, advertisements, or delivering potentially useful information. It is not necessary for everyone to have immediate interaction with robots. However, if we consider further deployments, we could see a situation in which everyone might need to use social robots. For instance, if a robot is used as an information provider (e.g., providing directions, as shown in Chapter 2.4), people would need to contact the robot to retrieve information. If a robot were used as a sales agent, then one would need to communicate with it if he/she would like to buy things in a shop.

In other domains, hesitation or negative attitudes have been reported. It is observed that people's opinions toward novel communication technologies tend to be highly polarized [1]. Thus, it would be plausible that people might have negative attitudes and emotions toward social robots because it is novel technological entity. As suggested in [2], it would be also plausible that people might avoid communicating with human-like social robots because they are the people who tend to avoid communicating with people.

This chapter is devoted to a study that addresses relationships between such negative attitudes in behavior and in people's attitude and anxiety. First, we report two psychological scales we have developed for measuring people's negative attitude and anxiety toward social robots. Then, an experiment was conducted in which participants engaged in simple interactions, including meeting, greeting, self-disclosure, and physical contact. Relationships among measured attitude and anxiety in psychological scales and behavior suggest how negative attitudes and anxiety affect their HRI behavior.

3.3.2 Psychological Scales for Human–Robot Interaction

We developed two psychological scales for *attitude* and *anxiety*. We briefly explain the idea behind these two scales. The detailed items of these scales are shown in the appendix.

---

* This chapter is a previously published paper (Nomura, Kanda, Suzuki, and Kato, 2008) in a modified version to fit with the context of this book.

3.3.2.1 Background: Attitudes and Anxiety

We consider the influence of *attitude* and *anxiety* in this study. These are psychological concepts that deal with different time scales. There is a precedent that involves these two concepts in studying people's behavior toward novel technological entities [3]. An *attitude* is defined as a relatively stable and enduring predisposition to behave or react in a certain way toward persons, objects, institutions, or issues; its source is cultural, familial, and personal [4].

Anxiety is defined as a feeling of mingled dread and apprehension about the future without a specific cause for the fear, a chronic fear of mild degree, strong overwhelming fear, a secondary drive involving an acquired avoidance response, or the inability to predict the future or to resolve problems [4]. Anxiety is generally classified into two categories: state and trait. The trait anxiety is the trend of anxiety as a stable characteristic in individuals, whereas the state anxiety is an anxiety transiently evoked in specific situations that change, depending on the situation and time [5].

3.3.2.2 Negative Attitudes toward Robots Scale (NARS)

The NARS was developed to determine human *attitudes* toward robots. This scale consists of 14 questionnaire items classified into three subscales: S1, "negative attitude toward interaction with robots" (six items); S2, "negative attitude toward the social influence of robots" (five items); and S3, "negative attitude toward emotional interactions with robots" (three items). Each item is scored on a five-point scale: (1) strongly disagree; (2) disagree; (3) undecided; (4) agree; (5) strongly agree. The internal consistency, factorial validity, and test–retest reliability of the scale have been confirmed through two tests on Japanese samples [2,6]. The English versions of items in the NARS were obtained using formal back-translation. Some social research, including some cross-cultural studies, has been conducted based on this scale [7,8].

3.3.2.3 Robot Anxiety Scale (RAS)

The RAS was developed to determine human anxiety toward robots evoked in real and imaginary HRI situations. In contrast with the NARS, this scale aims to measure state-like anxiety that may be evoked by robots.

The scale consists of 11 questionnaire items classified into three subscales: S1, "Anxiety toward communication capacity of robots" (three items); S2, "Anxiety toward behavioral characteristics of robots" (four items); and S3, "Anxiety toward discourse with robots" (four items). Each item is scored on a six-point scale: (1) I do not feel anxiety at all; (2) I hardly feel any anxiety; (3) I do not feel much anxiety; (4) I feel a little anxiety; (5) I feel quite anxious; and (6) I feel very anxious. An individual's score on each subscale is calculated by adding the scores of all items included in the subscale.

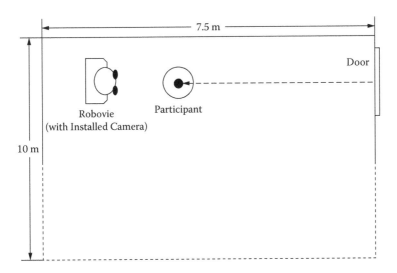

FIGURE 3.7
The room in which the experiment was conducted.

The internal consistency, factorial validity, and construct validity of the scale have been confirmed through the two tests for Japanese samples [9]. The English versions of the RAS items in the Appendix were also obtained through back-translation.

3.3.3 Experiment

The purpose of the experiment is to identify the relationships among negative attitudes, anxiety toward a robot, and behavior toward it.

3.3.3.1 Participants and Settings

There were 38 Japanese university students (22 males and 16 females) paid to participate, with a mean age of 21.3. A humanoid robot, Robovie II, was used. Participants interacted with the robot in a 10 × 7.5 m room shown in Figure 3.7.

3.3.3.2 Procedures

The procedures used in the experiment session were as follows.

1. Before meeting with the robot, participants filled presession questionnaires.
2. A participant was instructed to walk toward Robovie and greet it. Then, he/she entered the room alone. He/she walked to the front of Robovie.

3. After talking to Robovie for a preset amount of time (30 s), Robovie asked, "Tell me one thing that recently happened to you," to encourage participant response.

5. After they replied to Robovie or a set amount of time (30 s) passed, Robovie uttered a sentence to encourage physical contact: "Touch me."

6. After touching Robovie or a set amount of time (30 s) passed, the session was finished.

7. After the session, participants filled a postsession questionnaire.

Note that we intentionally designed the experiment to be short and simple in order to ensure that each participant's interaction with the robot was roughly the same.

3.3.3.3 Measurement

3.3.3.3.1 Anxiety and Negative Attitude

In the presession questionnaire, participants answered the NARS and RAS scale. They were asked to answer demographic information such as gender and age and other information about their personality. In addition, the *RAS* was administered in a postsession questionnaire to measure any change of participant anxiety toward robots due to the experimental session.

3.3.3.3.2 Communication Avoidance Behavior

We measured participants' communication avoidance behavior. Participants' behaviors were recorded using two digital video cameras. One of the cameras was installed in Robovie's eyes. The following items related to behaviors were extracted from the video data.

- *Distance (D)*: We measured the distance from participants to Robovie when first standing in front of it after entering the room. As distance is known as an index of subjective proximity in human communication [10] as well as in human–robot interaction [11,12,13], we considered that people who would hesitate to interact with the robot would need a longer distance.

- *Time before the first utterance (U1)*: We measured elapsed time before participants talked to Robovie after entering the room. This would indicate hesitation before participants voluntarily started to interact with the robot.

- *Time before the second utterance (U2)*: We measured elapsed time before participants replied to Robovie after being asked a question. In contrast to *U1*, this would indicate hesitation after participants requested to answer question from a robot.

- *Time before touching (T)*: We measured elapsed time before participants touched the robot's body after being encouraged to do so.
  While *U1* and *U2* concern hesitation before uttering, this *T* would
  indicate hesitation for making physical contact to the robot.

3.3.4 Results

3.3.4.1 Measurement Result

Figure 3.8 and Table 3.1 shows the means and standard deviations of the mea-
surements. Two-way mixed analysis of variances (ANOVAs) were conducted
to investigate the influence of gender (between participants) on RAS scores
measured before and after the experiment. There was a statistically significant
pre-/post-change in the S3 subscale scores. No significant pre-/post-change
was found in other subscales and no gender difference reported. The RAS-S3
(anxiety toward discourse with robots) score increased after the experiment,
indicating that participants got more anxious after experiencing interaction
with the robot. We consider that this indicate two things. First, the interaction
with the robot caused anxiety, probably because the interaction was designed
to be very short, which would make participants unsure whether their way of
communicating was appropriate or not. Second, RAS was designed to measure change in anxiety, which seems successfully achieved.

Student *t*-tests comparing male and female participants for the NARS
subscale scores and behavior indexes revealed no differences, except for
the S2 subscale score of the NARS, for which the difference was almost
statistically significant ($p = 0.054$), with a moderate effect size ($d = 0.665$).

(a) NARS

(b) RAS

(c) Behavior

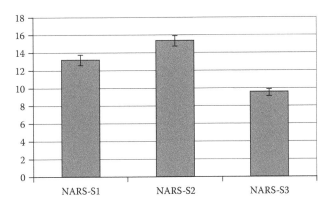

FIGURE 3.8
Measurement results.

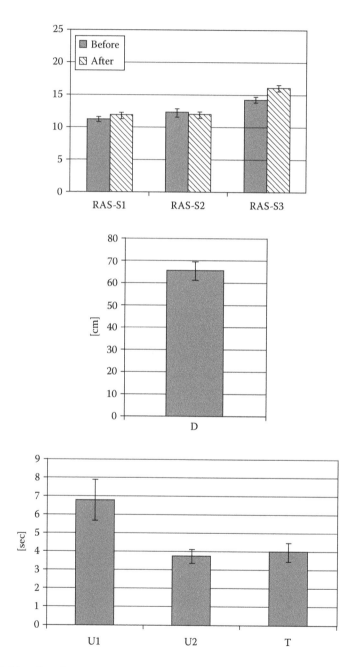

FIGURE 3.8 (continued)

TABLE 3.1

Mean and Standard Deviation of RAS Scores, NARS Scores, and Behavior Indexes, and Results of ANOVA and *t*-Tests

			Before Experiment		After Experiment		Gender			Before/After			Interaction		
		N	**Mean**	**SD**	**Mean**	**SD**	**F**	**p**	**partial η²**	**F**	**p**	**partial η²**	**F**	**p**	**partial η²**
RAS-S1	Male	21	11.9	2.5	12.4	2.3	2.589	.117	.069	2.740	.107	.073	.086	.771	.002
	Female	16	10.6	2.2	11.3	2.9									
RAS-S2	Male	21	12.0	3.7	11.6	3.6	.436	.513	.012	.364	.550	.010	.025	.875	.001
	Female	16	12.6	3.6	12.3	2.6									
RAS-S3	Male	21	14.2	3.1	16.1	3.7	.012	.912	.000	9.130	.005	.207	.033	.856	.001
	Female	16	14.4	2.4	16.1	2.6									
		N	**Mean**	**SD**	**t**	**p**	**d**			**N**	**Mean**	**SD**	**t**	**p**	**d**
NARS-S1	Male	22	13.0	3.1	−0.171	.865	.057	D(cm)		22	63.4	29.0	−0.519	.607	.176
	Female	15	13.3	4.3						15	68.0	21.9			
NARS-S2	Male	22	16.5	3.3	1.994	.054	.665	U2(sec)		22	3.6	1.6	−0.451	.665	.153
	Female	15	14.3	3.6						15	3.9	3.0			
NARS-S3	Male	22	9.8	2.3	0.711	.482	.237	T(sec)		15	3.6	2.5	−0.804	.429	.328
	Female	15	9.3	2.0						11	4.4	2.8			
	Male							U1(sec)		22	7.2	10.5	0.296	.769	.100
	Female									15	6.4	3.0			

3.3.4.2 Prediction of Communication Avoidance Behavior from Anxiety and Negative Attitudes

To investigate the relationship among *anxiety, negative attitudes*, and *communication avoidance behavior*, we conducted linear regression analyses. The NARS and RAS scores were used as independent variables and the behavior indexes as dependent variables. Thus, we aimed to identify a regression to predict *communication avoidance behavior* from NARS and RAS score. Previous experiments have suggested gender differences in these relations [2]. Thus, regressions were carried out for the complete sample, and the male and female subgroups separately.

Table 3.2 shows the statistically significant regression models in the analysis, revealing that the elapsed time before participants talked to Robovie

TABLE 3.2

Statistically Significant Regression Models Between NARS and RAS Scores and Behavior Indexes

Complete Sample							
Dependent Variable: U	β	*t*	*p*	**Dependent Variable: T**	β	*t*	*p*
RAS-S2	.186	1.140	.263	RAS-S1	.358	1.978	.061
RAS-S3	.387	2.385	.024	NARS-S2	−.549	−2.770	.011
NARS-S1	.227	1.375	.179	NARS-S3	.232	1.172	.254
NARS-S2	−.115	−.666	.511	R^2 (N = 26)	.241		
NARS-S3	−.069	−.418	.679				
R^2 (N = 37)	.218						

Male Subjects							
Dependent Variable: U	β	*t*	*p*	**Dependent Variable: T**	β	*t*	*p*
RAS-S1	−.242	−1.253	.231	RAS-S1	.468	1.949	.087
RAS-S2	.007	.032	.975	RAS-S2	.490	2.213	.058
RAS-S3	.530	2.492	.026	NARS-S1	−.377	−1.735	.121
NARS-S1	.496	2.671	.018	NARS-S2	−.461	−1.747	.119
NARS-S2	−.166	−.739	.472	NARS-S3	.445	1.998	.081
NARS-S3	−.055	−.295	.772	R2 (N = 15)	.555		
R^2 (N = 22)	.468						

Female Subjects							
Dependent Variable: D	β	*t*	*p*	**Dependent Variable: U**	β	*t*	*p*
RAS-S1	.589	2.550	.029	NARS-S3	.526	2.232	.044
NARS-S1	.522	2.196	.053	R^2 (N = 16)	.222		
NARS-S2	−.926	−3.578	.005				
NARS-S3	.209	.895	.392				
R^2 (N = 15)	.430						

after entering the room (*U*1) was influenced by anxiety toward discourse with robots (RAS–S3). The elapsed time before participants touched the robot after being encouraged (*T*) was influenced by the anxiety about communication capacity of robots (RAS–S1) and negative attitudes toward the social influence of robots (NARS–S2). RAS–S3 and RAS–S1 positively influenced *U*1 and *T*, indicating that more *anxiety* would cause more hesitation in their behavior. NARS–S2 negatively influenced T.

Gender analysis resulted in different models for males and females to predict behavior toward robots. For the male sample, *U*1 was positively influenced by RAS–S3 and negative attitudes toward interaction with robots (NARS–S1), and *T* was positively influenced by RAS–S1, anxiety toward behavioral characteristics of robots (RAS–S2), and negative attitudes toward emotional interaction with robots (NARS–S3). Moreover, *T* was negatively influenced by NARS–S2 with β coefficient –0.461, although this value was not statistically significant.

On the other hand, regression analysis for the female subgroup resulted in different models. For this group, distance from participants to Robovie when first standing in front of it after entering the room (*D*) was positively influenced by RAS–S1 and NARS–S1 and negatively by NARS–S2. NARS–S3 positively influenced the elapsed time before participants replied to Robovie after it encouraged participant responses (*U*2). Furthermore, the values of *R*2 showed that the regression models for *D*, *U1*, and *T* restricted to each gender had higher goodness-of-fit than those for the complete sample.

Overall, the results of the regression analysis revealed that the participants' anxiety and negative attitudes toward robots are related to their communication avoidance behavior toward robots, such as time spent in talking with and touching them. The result also indicates a gender difference in relationships between negative attitudes and anxiety, and behavior toward robots. The experimental results suggest that males with high negative attitudes and anxiety toward interaction with robots tend to avoid talking with them. In contrast, the results suggest that in case of females, those having anxiety toward interaction capacity of robots, high negative attitudes toward interaction with robots, and low negative attitudes toward the social influence of robots tended to avoid staying near robots. Further details of analysis is reported in [14].

3.3.5 Conclusion

We investigated why some people hesitate to interact with social robots. We focused on two psychological concept, *attitude* and *anxiety*, which are known to be influential to people's behavior. Two psychological scales were developed: the *Negative Attitudes toward Robots Scale (NARS)* and the *Robot Anxiety Scale (RAS)*. We conducted an experiment to reveal the relationships between negative attitudes, anxiety, and behavior toward robots, in which participants engaged in simple interaction with a humanoid robot, Robovie. The results revealed that negative attitudes and anxiety measured by these scales predicted participant communication avoidance behavior toward the robot. In addition, there is gender difference in the way negative attitude and anxiety influencing to communication avoidance behavior. These results imply that it is necessary in robotics design to take account of these relationships between anxiety, negative attitudes, and behavior toward robots.

Acknowledgments

The authors would like to thank Prof. K. Sugawara of the University of the Sacred Heart Tokyo and his colleagues of the Japanese Association of Psychology on Social Anxiety and Self for providing them with useful suggestions concerning their experimental results. They would also like to thank Dr. D. F. Glas of ATR Intelligent Robotics and Communication Laboratories for his collaboration with back-translation of the RAS.

References

1. Joinson, A. N. (2002), *Understanding the Psychology of Internet Behaviors: Virtual World, Real Lives.* Hampshire: Palgrave Macmillan (Japanese Ed. A. Miura et al. [2004] Kitaoji–Shobo).
2. Nomura, T., Kanda, T., and Suzuki, T. (2006), Experimental investigation into influence of negative attitudes toward robots on human–robot interaction, *AI Soc.*, 20(2), pp. 138–150.
3. Brosnan, M. (1998), *Technophobia: The Psychological Impact of Information Technology.* Evanston, IL: Routledge.
4. Chaplin, J. P., Ed. (1991), *Dictionary of Psychology*, 2nd ed. New York: Dell Pub Co.
5. Spielberger, C. D., Gorsuch, R. L., and Lushene, R. E. (1970), *Manual for the State—Trait Anxiety Inventory.* Palo Alto, CA: Counseling Psychologists Press.
6. Nomura, T., Suzuki, T., Kanda, T., and Kato, K. (2006), Measurement of negative attitudes toward robots, *Interact. Stud.*, 7(3), pp. 437–454.
7. Nomura, T., Suzuki, T., Kanda, T., and Kato, K. (2006), Altered attitudes of people toward robots: Investigation through the Negative Attitudes toward Robots Scale, in *Proc. AAAI–06 Workshop Hum. Implications Human–Robot Interact.*, pp. 29–35.
8. Bartneck, C., Suzuki, T., Kanda, T., and Nomura, T. (2007), The influence of people's culture and prior experiences with Aibo on their attitude towards robots, *AI Soc.*, 21(1/2), pp. 217–230.

9. Nomura, T., Suzuki, T., Kanda, T., and Kato, K. (2006). Measurement of anxiety toward robots, in *Proc. 15th IEEE Int. Symp. Robot Hum. Interact. Commun. (RO-MAN 2006)*, pp. 372–377.
10. Hall, E. T. (1966). *The Hidden Dimension*. Doubleday, Garden City, NY.
11. Kanda, T., Ishiguro, H., Imai, M., and Ono, T. (2003). *Body Movement Analysis of Human-Robot Interaction*. Paper presented at the International Joint Conf. on Artificial Intelligence (IJCAI2003).
12. Michalowski, M. P., Sabanovic, S., and Simmons, R. (2006). *A Spatial Model of Engagement for a Social Robot*. Paper presented at the IEEE International Workshop on Advanced Motion Control.
13. Walters, M. L., Dautenhahn, K., Boekhorst, R. T., Koay, K. L., Kaouri, C., Woods, S., et al. (2005). *The Influence of Subjects' Personality Traits on Predicting Comfortable Human-Robot Approach Distances*. Paper presented at the CogSci-2005 Workshop: Toward Social Mechanisms of Android Science.
14. Nomura, T., Kanda, T., Suzuki, T., and Kato, K. (2008). Prediction of human behavior in human-robot interaction using psychological scales for anxiety and negative attitudes toward robots. *IEEE Transactions on Robotics, 24*(2), 442–451.
15. Nomura, T., Suzuki, T., Kanda, T., and Kato, K. (2006). Measurement of negative attitudes toward robots. *Interaction Studies, 7*(3), 437–454.
16. Nomura, T., Suzuki, T., Kanda, T., and Kato, K. (2006). *Measurement of Anxiety toward Robots*. Paper presented at the IEEE Int. Symposium on Robot and Human Interactive Communication (RO-MAN2006).

Appendix—Items in the NARS [15] and RAS [16].

TABLE 3.3

Negative Attitudes toward Robots Scale and the Robot Anxiety Scale, With

Negative Attitudes toward Robots Scale (NARS)	
Subscale	**Item**
S1: Negative Attitude toward Interaction with Robots	I would feel uneasy if I were given a job where I had to use robots.
	The word "robot" means nothing to me.
	I would feel nervous operating a robot in front of other people.
	I would hate the idea that robots or artificial intelligences were making judgments about things.
	I would feel very nervous just standing in front of a robot.
	I would feel paranoid talking with a robot.
S2: Negative Attitude toward Social Influence of Robots	I would feel uneasy if robots really had emotions.
	Something bad might happen if robots developed into living begins.
	I feel that if I depend on robots too much, something bad might happen.
	I am concerned that robots would be a bad influence on children.
	I feel that in the future society will be dominated by robots.
S3: Negative Attitude toward Emotional Interactions with Robots	I would feel relaxed talking with robots.*
	If robots had emotions, I would be able to make friends with them.*
	I feel comforted being with robots that have emotions.*

Robot Anxiety Scale (RAS)	
Subscale	**Item**
S1: Anxiety toward Communication Capacity of Robots	Whether the robot might talk about irrelevant things in the middle of a conversation.
	Whether the robot might not be flexible in following the direction of our conversation.
	Whether the robot might not understand difficult conversation topics.
S2: Anxiety toward Behavioral Characteristics of Robots	What kind of movements the robot will make.
	What the robot is going to do.
	How strong the robot is.
	How fast the robot will move.
S3: Anxiety toward Discourse with Robots	How I should talk to the robot.
	How I should respond when the robot talks to me.
	Whether the robot will understand what I am talking about.
	Whether I will understand what the robot is talking about.

(* Reverse Coded Item)

4

Modeling Natural Behaviors for Human-Like Interaction with Robots

4.1 Introduction

What is natural behavior in human–robot interaction? If a seemingly human-like robot does not move in a human-like way at all, one would immediately find it "unnatural." Imagine if you were facing a human-like robot supposedly talking to you but its head direction is averted from you and facing toward someone else, you might well think that the robot is talking to the other person. Such nonverbal cues are known to be important in human communication, and recent studies have revealed that it is important to replicate nonverbal cues in human–robot interaction, too.

For instance, a robot that coordinates its body direction and exhibits a pointing gesture is successful in more comprehensive direction giving [1]. When a person is talking to a robot, it has been demonstrated that a robot that responds to a person's speech nonverbally provides an impression that it is listening to, understanding, and sharing information with the human speaker [2]. In a joint-attention task, it was revealed that people read a robot's nonverbal cues to infer a robot's internal state [3].

Moreover, detailed role of nonverbal behaviors have been revealed in human–robot interaction. Gaze has been successfully used to provide feedback [4], maintain engagement [5,6], and to adjust the conversation flow [7,8]. Pointing is another useful gesture to indicate a conversation's target objects [9,10].

This chapter deals with a couple of studies for modeling natural behavior for a human-like robot. In the first study, we introduce a work related to deictic interaction. When we talk about objects, we often engage in deictic interaction, using reference terms with pointing gestures. Such deictic interaction often happens when we start to talk about new things that are outside our shared attention [11]. When robots are operated in a public field, one of the salient natural interactive features is the use of deictic interaction. With such intuitive expression, people can quickly receive information from a robot. In our earlier work, we modeled the use of Japanese reference terms, *kore*, *sore*, and *are*, and learned about human behaviors in deictic interactions [12].

However, that did not make the interaction natural enough. In Section 4.2, we introduce a study to make deictic interaction natural.

In the second study, we introduce a work related to the phenomena known as *proxemics* [13]. During conversations, people adjust their distance to others based on their relationship and the situation. For example, personal conversations often happen within 0.45–1.2 minutes. When people talk about objects, they form an O-shaped space that surrounds the target object [14]. By doing so, each participant can look at the target object as well as the other people in the conversation. Such findings in human communication have been replicated in human–robot interaction. Since robots are mobile, a robot needs to control its position during interaction. Initial work conducted in the 1990s revealed the fact that obtaining control over a standing position enables a robot to stand in a line [15]. Recent studies have revealed that people are concerned about what they perceive to be appropriate distance between them and a robot [16]. In Section 4.3, we report a work in which we modeled the relationship between position and attention.

In the third study, we introduce a work for integrating nonverbal behavior for a direction-giving scenario in which timing comes to be important as well. Direction giving has often been a target of research in human communications as it naturally includes gestures in interaction. While such studies revealed how we use gestures [17], as well as efficient ways to convey route information [18], when we consider developing a robot for direction giving we found missing ways to integrate such knowledge into human communication studies. In Section 4.4, we report a work that addresses such integration related to nonverbal cues, in which a model of human listener comes to be highlighted.

The three studies included in the chapter are cases in which human-relevant behaviors are modeled and used by a human-like robot. Each model deals with a different aspect. The model introduced in Section 4.2 is build on a parametric model of deictic interaction, and it supplements aspect that were missing. The model in Section 4.3 is a case that human communication research might overlook because it concerns what might be considered trivial things for human communication. It does not involve a process related to language processing, thus is less cognitively demanding. However, this is not a trivial issue for a robot. What a human can intuitively do is not what a robot can easily do. The study demonstrates a case where we need a careful observation of humans' precise behaviors. The model in Section 4.4 concerns a case in which a robot does not simply mimic humans' behavior but instead behaved in a way to fit with a human model. While it would be easy to make a case for what is unnatural behavior, it is very complicated to make a behavior of a robot to be natural enough. Yet, we consider that establishing models is a path to make such a robot to fully behave in a natural way as humans do.

References

1. Ono, T., Imai, M., and Ishiguro, H. (2001). *A Model of Embodied Communications with Gestures between Humans and Robots*. Paper presented at the Annual Meeting of the Cognitive Science Society (CogSci2001).
2. Kanda, T., Kamasima, M., Imai, M., Ono, T., Sakamoto, D., Ishiguro, H. et al. (2007). A humanoid robot that pretends to listen to route guidance from a human. *Autonomous Robots, 22*(1), 87–100.
3. Breazeal, C., Kidd, C. D., Thomaz, A. L., Hoffman, G., and Berlin, M. (2005). *Effects of Nonverbal Communication on Efficiency and Robustness in Human-Robot Teamwork*. Paper presented at the IEEE/RSJ Int. Conf. on Intelligent Robots and Systems (IROS2005).
4. Nakano, Y. I., Reinstein, G., Stocky, T., and Cassell, J. (2003). *Towards a Model of Face-to-Face Grounding*. Paper presented at the Annual Meeting of the Association for Computational Linguistics (ACL 2003).
5. Rich, C., Ponsler, B., Holroyd, A., and Sidner, C. L. (2010). *Recognizing Engagement in Human-Robot Interaction*. Paper presented at the ACM/IEEE Int. Conf. on Human-Robot Interaction (HRI2010).
6. Sidner, C. L., Kidd, C. D., Lee, C., and Lesh, N. (2004). *Where to Look: A Study of Human-Robot Engagement*. Paper presented at the International Conference on Intelligent User Interfaces (IUI 2004).
7. Mutlu, B., Forlizzi, J., and Hodgins, J. (2006). *A Storytelling Robot: Modeling and Evaluation of Human-Like Gaze Behavior*. Paper presented at the IEEE-RAS Int. Conf. on Humanoid Robots (Humanoids'06).
8. Mutlu, B., Shiwa, T., Kanda, T., Ishiguro, H., and Hagita, N. (2009). *Footing in Human-Robot Conversations: How Robots Might Shape Participant Roles Using Gaze Cues*. Paper presented at the ACM/IEEE Int. Conf. on Human-Robot Interaction (HRI2009).
9. Kuzuoka, H., Oyama, S., Yamazaki, K., Suzuki, K., and Mitsuishi, M. (2000). *GestureMan: A Mobile Robot that Embodies a Remote Instructor's Actions*. Paper presented at the ACM Conference on Computer-supported cooperative work (CSCW2000).
10. Scassellati, B. (2002). Theory of mind for a humanoid robot. *Autonomous Robots, 12*(1), 13–24.
11. McNeill, D. (1987). *Psycholinguistics: A New Approach*. HarperCollins College Div.
12. Sugiyama, O., Kanda, T., Imai, M., Ishiguro, H., and Hagita, N. (2006). Humanlike conversation with gestures and verbal cues based on a three-layer attention-drawing model. *Connection Science, 18*(4), 379–402.
13. Hall, E. T. (1966). *The Hidden Dimension: Man's Use of Space in Public and Private*. The Bodley Head Ltd.
14. Kendon, A. (1990). *Conducting Interaction: Patterns of Behavior in Focused Encounters*. Cambridge University Press.
15. Nakauchi, Y., and Simmons, R. (2002). A social robot that stands in line. *Autonomous Robots, 12*(3), 313–324.
16. Dautenhahn, K., Walters, M. L., Woods, S., Koay, K. L., Nehaniv, C. L., Sisbot, E. A., et al. (2006). *How May I Serve You? A Robot Companion Approaching a Seated Person in a Helping Context*. Paper presented at the ACM/IEEE Int. Conf. on Human-Robot Interaction (HRI2006).

17. Allen, G. L. (2003). Gestures accompanying verbal route directions: Do they point to a new avenue for examining spatial representations? *Spatial Cognition and Computation, 3*(4), 259–268.
18. Daniel, M.-P., Tom, A., Manghi, E., and Denis, M. (2003). Testing the value of route directions through navigational performance. *Spatial Cognition and Computation, 3*(4), 269–289.

4.2 A Model of Natural Deictic Interaction[*]

Osamu Sugiyama, Takayuki Kanda, Michita Imai,
Hiroshi Ishiguro, and Norihiro Hagita

4.2.1 Introduction

When a robot operates in a public environment, we have observed a number of scenes in which *deictic interaction* is useful. When a robot provides route directions, it points at the direction of the destination. When it explains exhibits in a museum, the robot start explaining an exhibit by pointing at it. When a robot says "This one" while pointing, very quickly people comprehend which one the robot is referring to. As social robots are aimed to serve to ordinary people who do not have specialized computing and engineering knowledge, we consider that a *deictic interaction* enabled by human-like robots is one of the fundamental capability for social robots.

This chapter reports our work for modeling natural behavior for *deictic interaction*. In casual communication, people often use reference terms in combination with pointing, such as saying "Look at *this*" while pointing at an object in order to draw others' attention to the object. Reference terms are important for quickly and naturally informing the listener of an indicated object's location. The pointing gesture supplements the physical location of the target object referred to by the reference term.

One of the fundamental parts of such *deictic interaction* is the expression process and the recognition process. Basically, when a robot uses deictic expression, it is using a reference term with pointing. This process is called an *expression process*, for which we build a model that involves the way to use reference terms and pointing gestures [1]. While this addresses the case where the robot speaks with deictic expression while a partner human listens, we expect the opposite situation happens as well. The *recognition process* deals with such case, where we found the model for expression process can

[*] This chapter is a previously published paper, modified to be comprehensive and fit with the context of this book, by Sugiyama, O., Kanda, T., Imai, M., Ishiguro, H., and Hagita, N., 2007, Natural deictic communication with humanoid robots, *IEEE/RSJ Int. Conf. on Intelligent Robots and Systems (IROS2007)*, pp. 1441–1448.

be used for recognition as well [2]. Thus, these models enable a robot to perform the basic part of *deictic interaction*.

However, this simple functions only encompasses the processes related to the interpretation of information. We started to find that people consider such simple *deictic interaction* provided by the robot as not sufficiently natural. To supplement the missing part, we started to develop a process for facilitating interaction. This chapter reports on our modeling of the facilitation processes for *deictic interaction*, which are built on the expression and recognition processes.

4.2.2 Related Works

Deictic interaction has often been studied. One of the major study focuses was the recognition process: Methods were proposed to recognize the object by using a pointing gesture, an utterance, and relevant information [3,4,5]. Some studies integrated the recognition process with the expression process [6,7], though they address only static environment. In contrast, little study addressed the facilitating processes in *deictic interaction*.

4.2.3 Model of Deictic Interaction

A simple approach for *deictic interaction* involves the *interpretation process*, which includes expression and recognition in attention-drawing behavior. In addition, our proposal includes the *facilitation process*, which involves facilitating natural *deictic interaction* between humans and robots (Figure 4.1).

Interpretation Process
When we talk about the object in an environment, we point at the object and speak reference terms, such as 'this' and 'that', so that we draw the listener's attention quickly and intuitively to the object we consider in our mind. In this process a speaker would engage in *object indication* process, and a listener would engage in *object recognition* process.

Object Recognition
This is the process of interpreting an attention-drawing behavior based on its pointing posture as well as a reference term and object-property information in [8].

Object Indication
This is the process of generating a robot's attention-drawing behavior with pointing and a reference term in order to confirm to the recognized object with the interacting person.

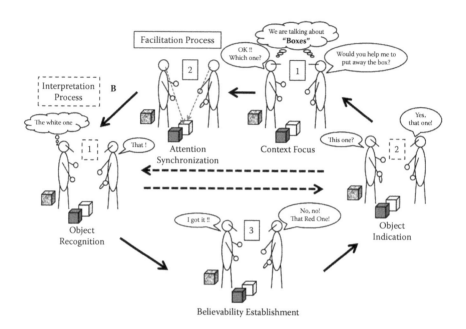

FIGURE 4.1
Interpretation and facilitation processes in natural deictic interaction.

Facilitation Process

While the *interpretation process* addresses the basic part of *deictic interaction*, we consider the needs of additional process to facilitate *deictic interaction*. When we talk about the object in an environment with attention-drawing behavior, the speaker always monitors the listener's reaction toward his behavior and evaluates whether the listener understands this behavior. At the same time, the listener also picks up the intention of the speaker's attention-drawing behavior and synchronizes his gaze by following the behavior. That is, *deictic interaction* is not a simple process where one person is pointing and another person is passively interpreting the pointing; it is a dynamic process in which the speaker and listener mutually engage in an interaction within a short time of a second or less. Without this dynamic process, *deictic interaction* would merely be giving a command to a static machine, which is not at all as natural as interhuman communication.

We believe that a robot should simultaneously react to human attention-drawing behavior as a listener while at the same time setting up an environment where the human as a listener feels comfortable in accepting the robot's attention-drawing behavior.

From this point of view, we added three new processes: *attention synchronization*, *context focus*, and *believability establishment*. Flow B in Figure 4.1 outlines the five processes to achieve natural *deictic interaction*.

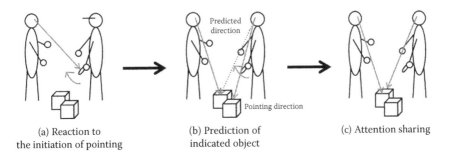

(a) Reaction to	(b) Prediction of	(c) Attention sharing
the initiation of pointing	indicated object	

FIGURE 4.2
The model of attention synchronization

Attention Synchronization

This process is to provide the interacting person with a feeling that the robot is attending to his or her pointing behavior. The flow of attention synchronization is shown in Figure 4.2. This consists of three subprocesses:

(a) Reaction to the initiation of pointing

When the listener notices the initiation of a speaker's attention-drawing behavior, the listener immediately starts to follow it with his gaze.

(b) Prediction of indicated object

Next, the listener estimates the intention of the speaker's attention-drawing behavior and predicts the object that the speaker is going to indicate. The listener often starts looking at the object before the pointing motion is finished.

(c) Attention sharing

Finally, when the speaker finished moving his hand in pointing, the speaker fixes the pointing gesture toward the indicated object and looks at it. At this time, the listener is also looking at the object. Thus, they are sharing their attention toward the object when the pointing motion is completed.

We took these actions as a reflex movement without noticing its importance in interhuman *deictic interaction*. However, attention synchronization must be an essential process in facilitating natural *deictic interaction*. Since such attention synchronization occurs in human communication, we believe that a robot should have this capability; otherwise, the speaker would have misgivings about the listener's understanding of the indicated object. In addition, we assume that attention synchronization makes the interaction quicker, since the speaker becomes more certain about the interaction.

Context Focus

This process sets up a presupposition toward the *deictic interaction* by utilizing attention-drawing behavior. In interhuman communication, the speaker and the listener share the objects of focus based on the context of the previous verbal communication before the *deictic interaction*. Without this process, there would be a flood of indication candidates in the living environment, resulting in the human feeling uncomfortable toward *deictic interaction* with a robot. By focusing on a certain kind of object, the human naturally immerses himself in *deictic interaction* with a robot.

Believability Establishment

This process provides human believability for a robot that is capable of interacting while using attention-drawing behavior. Communication, having ambiguous expressions such as those in d*eictic interaction*, is built upon the believability of the partner also being capable of managing an ambiguous expression; otherwise, the human would hesitate to interact with the robot in a deictic way. In this research, we construct this believability by creating a robot that is capable of error correction. In interhuman *deictic interaction*, the speaker can correct the listener's error immediately whenever he or she notices the listener's misunderstanding. It is very clear that a human has this ability, but it is not clear that a robot is capable of handling error correction. Therefore, it is essential to show that the robot has the ability to correct errors when it misunderstands a human's indication. Developing this ability in the robot is considered key to making humans possess believability toward the *deictic interaction* with the robot. Accordingly, this is one of the essential processes in achieving *deictic interaction* that is as natural as interhuman communication.

4.2.4 Development of a Communication Robot Capable of Natural Deictic Interaction

Based on the processes introduced in Section 4.2.3, we developed a communication robot that can engage in natural *deictic interaction* with a human. This section describes the details of this robot system.

Hardware Configuration

A robot system was developed using a communication robot, "Robovie" [9], a motion-capturing system, and a microphone. "Robovie" is 1.2 m tall with a 0.5 m radius and a human-like upper body designed for communicating with humans. It has a head (3 DOF), eyes (2*2 DOF), and arms (4*2 DOF). With a speaker in its head, it can produce sound output. With its 4-DOF arms, it can point with a gesture similar to that of humans. Using its motion-capturing system, we can obtain 3D positions from markers attached to the person, Robovie, and objects; accordingly, by using an Ethernet, the system obtains

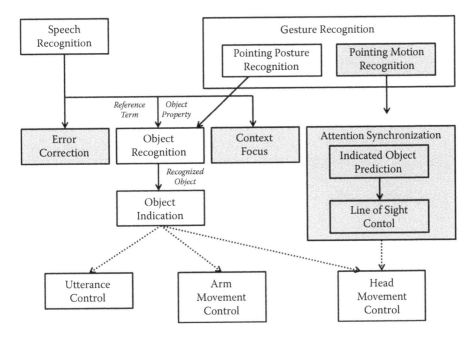

FIGURE 4.3
System configuration

its 3D positions as input and then calculates their locations. In addition, the speaker wears a microphone to avoid difficulty with speech recognition.

Software Configuration

The configuration of the developed robot system is shown in Figure 4.3. The interpretation processes in Figure 4.1 are achieved by the white-colored components. Meanwhile, the facilitation processes are achieved by the orange-colored components. Each process gets its input from either/both speech recognition or/and gesture recognition, and it controls Robovie's behavior as its output. For speech recognition input, we used a speech recognition system [10] and attached a microphone to a human in order to avoid noise (we appreciate that there are many studies on noise-robust speech recognition that do not require microphones attached to people). The speech recognition system can recognize reference terms, object-color information, and error-alerting words in human utterances. On the other hand, gesture recognition is done with a motion-capturing system. In gesture recognition, the system handles two types of processes: pointing posture recognition and pointing motion detection. The following describes the details of implementing each process in Figure 4.1.

Interpretation Processes   In [11], we developed a three-layered model for *interpretation processes*. We built a *reference-term model* by modeling the use of

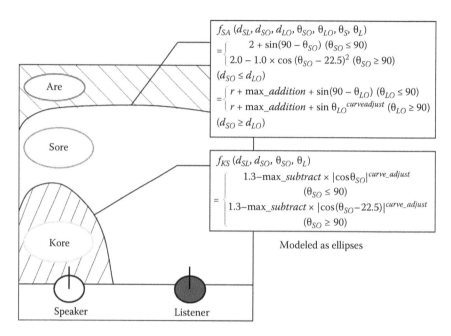

FIGURE 4.4
Reference-term and pointing space models

Japanese reference terms *kore, sore,* and *are,* and learned from human behaviors in deictic interactions. Data collection was conducted with 10 pairs of subjects who performed deictic interactions at various spatial configurations. As the analysis of the interaction, the boundary shape is represented as a function of the speaker–listener distance, body orientation (e.g., faced or aligned), and the object's location (Figure 4.4, left, shows a brief summary of this reference-term model).

In addition, the *pointing-space model* is prepared, which can specify the target modeled as the pointing space model (Figure 4.4, right). When the pointing does not specify the target, the *object–property model* is used so that the robot adds an adjective to further specify the object.

The interpretation processes are developed based on the above three-layer model. In object recognition, the system receives a reference term and object-property information from the speech recognition module, and pointing gesture information from gesture recognition in order to identify the indicated object. The details are described in [2]. In object indication, the system generates an appropriate attention-drawing behavior enabling the humanoid robot to confirm the human-indicated object with the same expression. The details are described in [1].

Facilitation Processes   This section describes the implementation of *facilitation processes* designed to facilitate natural *deictic interaction* between humanoid robots and humans. The following describes the details of the facilitation processes.

Context Focus   The *context focus* process is implemented as voice communication with the human. In the first stage of the interaction in our experiment, the robot (R) and the human (H) have a short conversation:

R: May I help you?
H: Yes, please help me to put away these *round boxes*.
R: OK. Would you tell me the order of the *boxes* to be put away?

In this case, the robot and the human can focus on the round boxes in the environment based on the context described above. It is very important for both robots and humans to be aware of specific objects indicated by the context. Otherwise, they would have to consider objects from a wide variety of candidates when they engage in deictic interaction. In this research, the robot could only handle the round boxes. Thus, this conversation is only for the human to mentally prepare for the objects in the context. Enabling robots to distinguish objects by context remains a future work.

Attention Synchronization

The process of attention synchronization is implemented by three sub-processes: *pointing motion detection, indicated object prediction*, and *line of sight control*. The important requirement for attention synchronization is that the robot starts looking at the indicated object quickly before the person finishes the pointing motion. Thus, the robot has to predict the indicated object from the pointing motion. The following describes each sub-process in order.

Pointing Motion Detection   This process monitors human arm motion and determines whether the present human motion is a pointing gesture. Human pointing gesture has rich patterns: some people twist the body to point at a left-side object with the right hand, and others first raise a finger near the head and then point at the object. On the other hand, humans also exhibit various movements when they are not pointing but just standing and talking. In order to detect a human pointing gesture in various situations, we utilize three different parameters for pointing gestures:

i) The speed of fingers v_p;
ii) The distance between the human's finger and body d_p;
iii) The angle between the human's vertical axis and his/her pointing direction θ_p.

If the human is in the process of performing a pointing gesture, one of these parameters should be increased. We analyze various pointing motions and set up a threshold for each parameter. The evaluation equation is given as follows.

$$f(t) = \begin{cases} 1 & \left(v_p > 100_{mm/s} \middle\| d_p > 400mm \middle\| \theta_p > 15°\right) \\ 0 & \left(v_p \leq 100_{mm/s} \middle\| d_p \leq 400mm \middle\| \theta_p \leq 15°\right) \end{cases}$$

where t is a frame of the motion-capturing data, and 1 means detection of pointing while 0 means no detection.

Pointing Object Prediction   This process predicts the object that a human is going to indicate in the first stage of pointing motion so that the robot can synchronize its gaze to the human's pointing motion. In order to predict the object that the human is going to point, the system predicts a future human finger position based on the present speed vector of the fingers. Utilizing the position of fingers, it searches for the indicted object.

The difficulty is in deciding the finger position at how many seconds later the system should use in predicting an indicated object. If this time interval is too long, the robot's line of sight goes past the indicated object and sees a farther object. If it is too short, the robot's line of sight stays on a nearer object. In order to solve this trade-off problem, we conducted an easy experiment to evaluate how far, in seconds after a finger starts the pointing motion, it will finish the end of its intended reach to indicate the object. The results led us to set a prediction time of 0.3 seconds later. The system predicts the human's final finger position at 0.1–0.3 seconds later, and based on this, it searches for the object and decides on the object nearest the calculated position temporally and spatially.

Line of Sight Control   This process controls Robovie's line of sight expressed by its head sliding toward the indicated object predicted by the process of pointing object prediction. The process controls Robovie's line of sight as follows;

1. If pointing gesture is not detected, the process controls the line of sight to see the human's face.
2. If pointing gesture is detected and indicated object is predicted, the process controls the line of sight to see the indicated object.
3. If pointing gesture is detected but there are no predicted objects, the process controls the line of sight to see the former predicted object (or see the human's face, if there are no predicted objects in the past).

Error Correction

The process of error correction is to establish believability toward the robot's believability. It is implemented as voice communication with the human.

If the process of speech recognition detects error-alerting words in a human utterance, this process is called by the system. In this process, the robot asks the human to repeat his or her indication and deletes the latest recognized object from its memory. By successfully using this process on the robot system, the human can feel that it is possible to correct the indication if the robot initially misunderstands it. Consequently, this improves the human's sense of believability toward *deictic interaction* with the robot.

4.2.5 Evaluation Experiment

We conducted an evaluation experiment to reveal the effectiveness of the facilitation process. We compared the developed system with a baseline system. In the baseline system, participants interacted in a nondeictic way when asked to tell the unique name of the object. We set up a simplified situation in which *deictic interaction* could be used. In the room, participants were asked to refer the boxes to the robot. We measured how participants subjectively perceived the interaction as natural and effective.

4.2.5.1 Method

Experimental environment. The experiment was conducted in a rectangular room 7.5 m × 10.0 m (Figure 4.5). Because of the limitations imposed by the motion-capturing system, we only used an area 3.5 m × 3.5 m in the center of the room. There are five cylindrical boxes (radius 110 mm, height 300 mm). At the beginning of each session, each participant was asked to place these five boxes freely in the area except for places to where Robovie could not

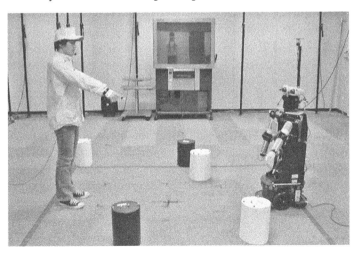

FIGURE 4.5
A scene from the experiment.

point. The participant stood in front of the robot, usually at distance of 0.5 m to 1.5 m from the robot.

    Participants. Thirty university students (16 men, 14 women) participated in the experiment.

    Conditions. We adopted the 2 (facilitation process) × 2 (method of instruction) conditions for the experiments. The definition of each factor is as follows.

Facilitation Process Factors

With facilitation process (a).   Before the session, participants were instructed how to correct errors of the robot, and they could do so freely as they gave orders (error correction). They were also asked to start giving an order while explicitly mentioning the context. For example, they might say, "Please bring the **boxes**" (context). During the session, the robot controlled its gazing direction to the direction in which the participants were going to point, which is based on the developed mechanism of attention synchronization (attention synchronization). Note that the robot also looked at boxes at the confirmation phase.

Without facilitation process (a′).   The participants were given none of the instructions mentioned above. When participants gave the object order, the robot did not control its gazing direction but just looked in the participant's direction. Participants were not allowed to correct errors of the robot. Note that the robot looked at boxes at the confirmation phase.

Method of Instruction Factor

    (b) *Deictic method* (pointing + reference term). Participants were asked to use pointing gesture, reference terms and object color information when they give orders to the robot.

    (b′) *Symbolic method.* Participants were asked to read the numbers on the boxes when they gave orders to the robot. Two-digit IDs were attached to the box in 14-point font, which is readable from a 2-m distance.

This setup is intended to simulate a situation where there is some difficulty in finding the symbol used to identify the object. In our daily situations, for example, we might say, "Please look at the book entitled *Communication Robots.*" Here, we need to see the characters on the object and read them. This load was simulated with the two-digit ID with a font size of 14 points.

    The experiment was a within-participant design, and the order of experimental trials was counterbalanced.

4.2.5.2 Procedures

Before the experiment, participants were instructed on how to interact with the robot, by both the deictic method (reference terms and pointing) and the symbolic method. After the instructions, participants experienced four sessions for all four conditions. In each session, they conducted the following procedures three times:

1. They freely place the five boxes.
2. Under (a) condition, they talk about context.
3. They decide the order of five boxes.
4. They indicate the five boxes one by one. The method of instruction is either (b) or (b'), depending on the experimental condition. For example, under (b) condition, a participant might point at the first box and say "This" and then at the second box and say "That white one." The participant continues this until the fifth object.
5. After the participants indicate the fifth box, Robovie repeats the given order. This time, the robot looks at each object with its utterance for confirmation. For example, under (b) condition, Robovie uses reference terms while pointing at each of the five boxes.

After the three repetitions, participants fill out a questionnaire to give their impressions of the interaction with the robot. We designed the interaction to include the quickness of indicating objects; for this purpose, participants continuously indicate a box five times. Since the impression of the interaction could depend on the placement of the boxes, we gave participants three opportunities to experience the interaction under each condition.

4.2.5.3 Measurement

We expected that the facilitation process would make the *deictic interaction* with the robot more natural. In addition, we expected that the attention-synchronization mechanism would give a better impression of sharing spatial information of the boxes. Thus, we measured the following impressions with the questionnaire. The participants answered each question on a 1-to-7 scale, where 1 stands for the lowest evaluation and 7 stands for the highest.

Naturalness: Your feelings about the naturalness of the conversation

Sharing information: Your feelings of sharing information with the robot

Regarding the comparison between the deictic method and the symbolic method, we are interested in quickness and correctness [12]. Thus, we also measured the following impressions in the questionnaire:

Quickness: Quickness of your indication of the boxes

Correctness: Correctness of the robot's identifying your indication of the boxes

Understanding: The robot's understanding of the indication

We also measured the system's performance as the accuracy of recognizing indication in both the deictic method and symbolic method. The main cause of error comes from failure in speech recognition. Since speech recognition sometimes fails whereas the deictic method utilizes multimodal input, we expected the deictic method to provide better performance in reality.

Performance: The rate that the robot system correctly identified the object indicated by the participants

4.2.5.4 Results

Figures 4.6-1 and 4.6-2 show the results of the questionnaire. We conducted repeated measures of the ANOVA (analysis of variance). There are two within-participant factors, *facilitation process* (with facilitation process or without facilitation process) and *method of instruction* (deictic method or symbolic method).

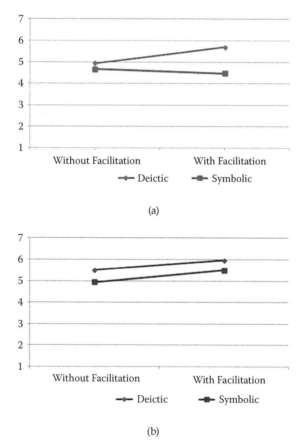

(a)

(b)

FIGURE 4.6-1
Results for facilitation process: (a) naturalness, (b) sharing information.

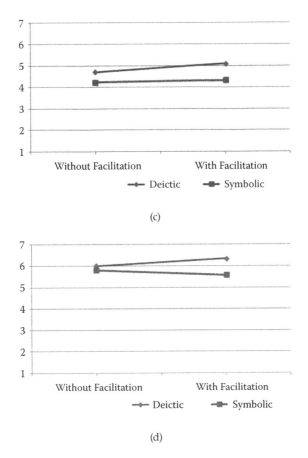

FIGURE 4.6-2
Results for method of instruction: (c) quickness, (d) robot's understanding.

Naturalness (Figure 4.6-1a) There were a significant difference in the method of instruction (F[1,29] = 9.715, p = .004) and the interaction between the two factors (F[1,29] = 8.745, p = .006). Analysis of the interaction resulted in a significant difference in the method of instruction under the with-facilitation process condition (p < .01), as well as a significant difference between the facilitation processes under the *deictic interaction* condition (p < .01). Thus, the facilitation process contributed to the feeling of natural interaction when participants interacted with the robot in *deictic interaction*.

Sharing information (Figure 4.6-1b): There were significant differences the method of instruction (F[1,29] = 11.197, p = .002) and in the facilitation process (F[1,29] = 4.969, p = 0.34). There is no significant difference in the interaction between the two factors. Thus, participants felt that the robot with a facilitation process is better at sharing information than a robot without it. In addition, it was revealed that *deictic interaction* contributed to the feeling of sharing information better than could symbolic communication.

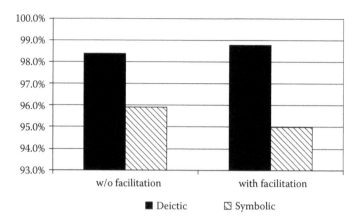

FIGURE 4.7
Performance.

Quickness (Figure 4.6-2c): There were significant differences related to the instruction method (F[1,29] = 8.629, *p* = .006). There was no significant difference in the facilitation process, interaction between the two factors, and performance. Thus, participants felt that the deictic method is quicker than the symbolic method.

Correctness: There was no significant difference in facilitation process, instruction method, and interaction between the two. Note that the correctness has a significant correlation with the performance (Pearson correlation = .318, *p* < .01). This suggests that participants might associate the robot's real performance as the correctness impression, but they did not feel the difference in either facilitation process or instruction method.

Understanding (Figure 4.6-2d): There was significant difference in the performance (F[1,29] = 10.563, *p* = .003), and no significant difference in the other factors. Thus, participants distinguished the real performance and evaluated it as the robot's understanding.

Performance (Figure 4.7): For performance, we conducted within-participant design ANOVA, which indicates a significant difference in the method of instruction (F(1,29) = 18.448, *p* < .001). There was no significant difference in the facilitation process. Thus, the multimodal input used in the deictic method positively affected the system performance.

4.2.5.5 Summary

The result revealed the effectiveness of the deictic interaction.

As the deictic method merits from multiple source of information (pointing and reference term), the robot's performance in recognition is better when people used deictic impression. Further, participants perceived it quicker when they used deictic expression. These findings reveals the efficiency in using deictic interaction.

Moreover, the result revealed the effect of the facilitation process. When accompanied with the facilitation process, the robot's deictic interaction provides natural impression. In contrast, without facilitation process, they perceived the interaction less natural and not different with the alternative method. We believe that the facilitation process is the key process to make the deictic interaction natural.

4.2.5.6 Limitations

As the state of the study is at the level of providing proof-of-concept, there are limitations to making the model generally available. As reference terms are language dependent, apparently we need different reference term models for different languages. The facilitation process is implemented in a naive way, thus would need adjustment depending on the context. Believability establishment would be largely alternated if the robot has different expectation because it will be either widely used across society or used repeatedly. In such case, perhaps robots would not need to establish believability at each interaction.

4.2.6 Conclusion

This chapter reports the model for *deictic interaction*. It is based on the *interpretation process*, which address expression and recognition with pointing gestures and reference terms. In addition, there are *facilitation processes* added, which consist of three subcomponents: attention synchronization, context focus, and believability establishment. The model was implemented in a humanoid robot, Robovie, with a motion-capturing system. In the experiment, it was tested whether people prefer to engage in *deictic interaction* or baseline method. The results support the needs of *facilitation process*. Only with the *facilitation process deictic interaction* is perceived as more natural than the baseline method, but not only with the *interpretation process*. This study reveals the challenge of effectively using the human-like body property of robots, which does not only concern the aspect of information exchange, but requires more dedicated observation of subtle aspects of human communication.

Acknowledgments

This research was supported by the Ministry of Internal Affairs and Communications of Japan.

References

1. Sugiyama, O., Kanda, T., Imai, M., Ishiguro, H., and Hagita, N. (2005). Three-layered draw-attention model for the humanoid robots with gestures and verbal cues, *IEEE/RSJ Int. Conf. on Intelligent Robots and Systems (IROS2005)*.
2. Sugiyama, O., Kanda, T., Imai, M., Ishiguro, H. and Hagita, N. (2006). Three-layer model for generation and recognition of attention-drawing behavior, *IEEE/RSJ Int. Conf. on Intelligent Robots and Systems (IROS2006)*, pp. 5843–5850.
3. Haasch, A., Hofemann, N., Fritsch, J., and Sagerer, G. (2005). A multi-modal object attention system for a mobile robot, *Int. Conf. on Intelligent Robots and Systems (IROS2005)*, pp. 1499–1504.
4. Hanafiah, Z. M., Yamazaki, C., Nakamura, A., and Kuno, Y. (2004). Understanding inexplicit utterances using vision for helper robots, *Int. Conf. on Pattern Recognition*.
5. Inamura, T., Inaba, M., and Inoue, H. (2004). PEXIS: Probabilistic Experience Representation Based Adaptive Interaction System for personal robots, *Systems and Computers in Japan*, Vol. 35, No. 6, pp. 98–109.
6. Scassellati, B. (2000). Investigating models of social development using a humanoid robot, *Biorobotics*, MIT Press.
7. Breazeal, C., Kidd, C. D., Thomaz, A. L., Hoffman, G., and Berlin, M. (2005). Effects of nonverbal communication on efficiency and robustness in human-robot teamwork, *Int. Conf. on Intelligent Robots and Systems (IROS2005)*, pp. 383–388.
8. Sugiyama, O., Kanda, T., Imai, M., Ishiguro, H., and Hagita, N. (2006). Three-layer model for generation and recognition of attention-drawing behavior, *Int. Conf. on Intelligent Robots and Systems (IROS2006)*.
9. Kanda, T., Ishiguro, H., Imai, M., and Ono, T. (2004). Development and evaluation of interactive humanoid robots, *Proceedings of the IEEE*, Vol. 92, No. 11, pp. 1839–1850.
10. Ishi, C. T., Matsuda, S., Kanda, T., Jitsuhiro, T., Ishiguro, H., Nakamura, S., and Hagita, N. (2006). Robust speech recognition system for communication robots in real environments, *Int. Conf. on Humanoid Robots (Humanoids2006)*.
11. Sugiyama, O., Kanda, T., Imai, M., Ishiguro, H., and Hagita, N. (2006). Humanlike conversation with gestures and verbal cues based on a three-layer attention-drawing model, *Connection Science*, vol. 18, pp. 379–402.
12. Sugiyama, O., Kanda, T., Imai, M., Ishiguro, H., and Hagita, N. (2005). Three-layered Draw-Attention Model for Humanoid Robots with Gestures and Verbal Cues," *Int. Conf. on Intelligent Robots and Systems (IROS2005)*, pp. 2140–2145.

4.3 A Model of Proximic Behavior for Being Together While Sharing Attention

Fumitaka Yamaoka,[1,2] Takayuki Kanda,[1]
Hiroshi Ishiguro,[1,2] and Norihiro Hagita[1]

[1]*ATR IRC Laboratories,* [2]*Osaka University*

4.3.1 Introduction*

Our natural behavior concerns our reaction to scenes while walking around, too. For instance, when we visit shops and museums with friends, we often see exhibits and products together. While walking around, even without talking, we know what friend might be interested. We sometimes see the same object, and when a friend moves to the next one, we might move to it as well. Particularly when two people are together considering one or more objects, interaction is very dynamic, and we keep transiting our attention together from one object to another.

Since a robot is typically "mobile," desired "natural" behavior involves a way to control a robot's position during interaction. There is some early work in this area. For instance, in our previous work, we built a model for a robot to appropriately position itself while presenting information on products [1]. However, imagine the scene described in the beginning. The difficulty is not only the appropriateness of position, but the appropriateness of timing. When one shifts his/her attention to another target, in order to provide a feeling of "being together," an accompanying robot should move in timely manner. In the previous work, a robot was able to move only after the partner finished moving to the other object. This seems too late to cause a feeling of "being together." This chapter addresses a study to make it possible for a robot to replicate this sense. We started from analyzing people's behavior, and establish a model to be implemented in a robot. Finally, the model was tested in a experiment in which the developed robot interact with human participants autonomously.

4.3.2 Related Work

Previous research in human–robot interaction has looked at aspects of joint-attention [2]. Joint-attention between people is established using both implicit and explicit cues. An example of implicit cues is gazing toward an object, which is understood by others and responded to by their shifting attention towards the same object. In human–robot interaction, the role of gaze has been revealed through research in joint-attention (e.g., References 3, 4, and 5). Further, it is revealed that a subtle difference in gaze motion implicitly leaks information about one's attention [6].

Explicit cues are those also known as "deictic" references [7]. Examples of these references are gaze and pointing behavior accompanied by verbal references. Robots have also used the deictic gestures to draw others' attention to information in the environment [8]. Breazeal et al., who used the term

* This chapter is a modified version of a previously published paper, Yamaoka, F., Kanda, T., Ishiguro, H., and Hagita, N. (2009). *Developing a Model of Robot Behavior to Identify and Appropriately Respond to Implicit Attention-Shifting*. Paper presented at the ACM/IEEE Int. Conf. on Human-Robot Interaction (HRI2009) to be comprehensive and fit with the context of this book.

"implicit" and "explicit" to describe these cues, suggested that "implicit" cues are particularly important in communicating internal states and can significantly improve human–robot interaction [9].

However, previous work did not take into account aspects of mobility and proximity of the robot as cues that play a role in establishing joint attention. In human communication research, the importance of mobility is well recognized. Kendon suggested that when people shift attention, they move their gaze. If they continue to attend to their new focus of attention, they change their body orientation and form a space where people's attention focus together. This is known as the O-space [10]. As Kendon's study suggested, position and body orientation is used for sharing continuous and prolonged attention, while gazing and pointing is used to communicate instant attention.

Aspects of mobility and proximity have been studied in human–robot interaction. However, these studies did not focus on attention. A number of studies looked at trajectories of robot movement that keep a comfortable distance from people [11,12]. Further, Sisbot et al. developed a path-planning algorithm that takes into account people's positions and orientations in order not to disturb them [13]. Pacchierotti et al. studied passing behavior and developed a robot that waits aside to make a allow people to pass by [14]. Gockley et al. found that following the direction of a person (instead of the person's trajectory) is a good way for a robot to follow a person [15] and create feelings of "being together." However, these studies focused only aspects of proximity and robot movement. How these aspects might relate to shifting of attention remains unexplored.

4.3.3 Modeling of Implicit Cue for Attention-Shifting

4.3.3.1 Definition of "Implicit" Attention-Shifting

In the study we address interaction in which two people look around together. Figure 4.8 shows one of such situations where we collected data on the interaction of two persons. In the setting of the study, they watch posters in the room together. When they finish reading information, they move to other posters together. This imitates typical situations where we go shops and museum together with friends.

In some case, they would share attention through an explicit cue. For instance, one would verbally ask another, "Let's move to the next one," perhaps in combination with pointing to the next object. Such an explicit cue is so obvious that one should be expected to respond at the moment when such an explicit cue is exhibited. Thus, it is relatively easy to develop a model for a robot to timely respond.

In contrast, we have observed that people often do this joint movement interaction in silence. When one moves to another object, the partner immediately follows. This enables them to behave in a way that they feel they are "being together." As there is no explicit cue, it is not mandatory to follow

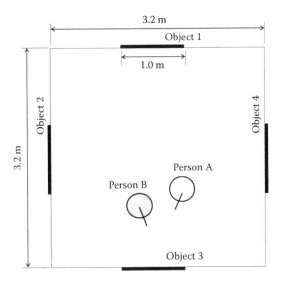

FIGURE 4.8
Experimental situation: (a) pay attention to one object, (b) pay attention to two objects.

in a timely manner. However, continuous failure in following may result in providing an impression that one is not behaving like "being together."

4.3.3.2 Overview of Modeling

For modeling, we observed dyads establishing joint attention at, and talking about, objects placed in different spots in a room (Figure 4.8). Five pairs of undergraduate students (seven men and three women with an average age of 21) participated. In the room, four panels displaying cell phone information were placed in different spots (Figure 4.8). Participants were asked to compare cell phones and rank each one from the highest to the lowest price. We recorded participants' behaviors with a video camera.

In general, we observed that people frequently share attention in implicit way. In our study, we did not require a particular way for them to conduct tasks, thus, they could use explicit cues. However, their natural behavior is to use implicit cues frequently. This matches our initial view that people often share attention in implicit ways.

Thus, we further analyzed how people share attention with implicit cues. Our modeling is based on analysis of people's interaction, as well as being informed by the study conducted by Kendon [16]. Kendon suggested that body position and orientation are the most important implicit cues that indicate people's attention. Thus, in the analysis, we focused on how people use these implicit cues. We analyzed three aspects of interaction, which turned into three submethods of the proposed model.

4.3.3.3 Estimating Attention from Implicit Cues

The first analysis concerns the estimation of attention from implicit cues, aiming to answer the question of how people know the object of their partner's attention. Here, we referred Kendon's work [16], reporting that humans can make transactional segments to their final goal of attention, viewing, talking about, and touching them. The transactional segment is made in front of a human and the width of it is changed according to the spatial arrangement of the surroundings of the person such as tables or walls and so on. Here, we defined the transactional segment as the space in front of a person because there is no obstacle between him/her and the object in an experimental setting.

Our observations revealed that there are two main patterns of spatial arrangements between participants and objects when they establish shared attention of objects. In the cases where they pay attention to one object, there is only the object in their transactional segment (Figure 4.9a). In cases where they pay attention to two objects, both objects are in their transactional segment (Figure 4.9b). To replicate such human's interaction, we define the following method for estimating a partner's attention target.

Method of estimating attention from implicit cues: Partner's attention can be estimated by identifying whether each object is in partner's transactional segment.

4.3.3.4 Detecting Partner's Attention-Shift from Implicit Cues

The second analysis concerns detection of the moment when a partner's attention shifts. Kendon reported that when people shift attention, they shift their gaze. For prolonged shift of attention, they reorient their body by the arranging their legs. This certainly happened in our case, too. In the observation, participants changed their body orientation to shift their attention. Thus, we define our method as:

Method of detecting implicit attention-shift: The change in the arrangement of a partner's legs notifies the shift of his/her attention.

Unfortunately, our current system cannot detect the position of people's legs. In this study, we assume partner's legs would be moved when body location moved over 50 cm, as this is beyond a typical size of human's stride.

4.3.3.5 Predicting Partner's Next Attention from Implicit Cues

The third analysis concerns the prediction of the partner's next attention before the partner stops to watch the next object. This prediction is necessarily if a robot aims to move to the next position in a timely manner, that is, before the partner arrives and stops at the next object. The question is which cues are useful for the prediction?

We coded the following items for each person based on our video data. In addition, we checked when each item begins and ends.

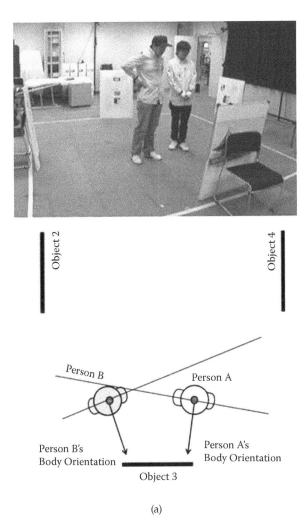

(a)

FIGURE 4.9
Estimating partner's attention targets.

- *Attention state*: Which object did the dyad talk about?
- *Gaze*: Which object did the person look at?
- *Turn*: Which object did the person finally turn toward?
- *Approach*: Which object did the person finally approach?

Figure 4.10 shows a sample from our coding. In this sample, initially participants talked about object 3. Then, the person B began to turn to object 1 and approached it. A also turned and approached the object after a short time. Finally, both A and B completed their approach to object 1 and began to talk about it.

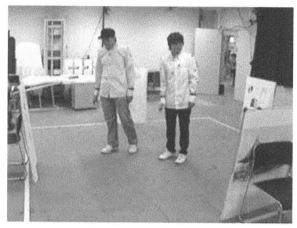

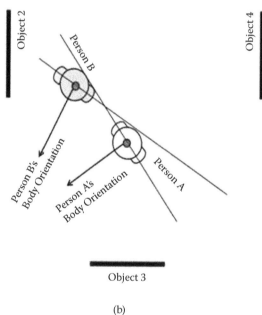

(b)

FIGURE 4.9 (continued)

It is well known that gaze is a cue to inform the focus of attention [17]. Our coding showed that the gaze direction is a useful cue to predict the object people will stop at and observe next. In the scene in Figure 4.10, four seconds in advance of person A's stopping at object 1 his gaze started to be directed to object 1. Thus, this means that the robot can approach the next target simultaneously with the partner's motion by checking its partner's gazing direction. We proposed the following method for predicting a partner's implicit direction of attention: *A partner's next attention can be predicted by checking that partner's gazing direction.*

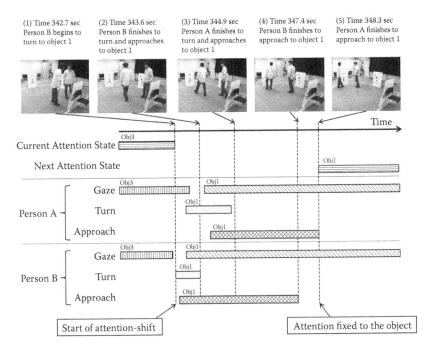

FIGURE 4.10
Example coding of our behavior data (participants' attention changes from object 3 to object 1).

4.3.4 Modeling of Spatial Position

In our previous work [1], we developed a position model for an information-presenting robot that establishes O-space. However, our previous model cannot deal with the case in which a partner pays attention to multiple targets. Thus, we updated our position model based on our observation.

Our previous model consisted of four constraints for establishing O-space: (1) proximity to partner, (2) proximity to object, (3) partner's field of view, and (4) robot's field of view [1]. To deal with the case that the robot needs to pay attention to multiple targets, we updated these constraints.

4.3.4.1 Constraint of Proximity

Our previous research showed the importance of the constraint of proximity to the partner and each object. The result from this time observation also supports this constraint. Participants tend to stand close to the partner and each object. Thus, we defined the constraint of proximity: *The robot should stand close to the partner and each object* (Figure 4.11).

4.3.4.2 Body Orientation

Our observation also informed our model regarding the spatial relationship that people form with each other and with the object to which they are

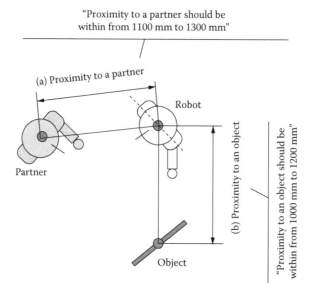

FIGURE 4.11
Constraint of proximity.

attending when they talk about this object. We found that a person's body orientation is close to the vector from the person to the object of his/her attention target when they pay attention to one target. We also found that a person's body orientation is close to the median vector between the two vectors from the person to each attention target when they pay attention to two targets (Figure 4.12). Thus, we defined the robot's body orientation as shown below.

Case 1: Partner pays attention to one target

The robot's body orientation follows the vector from robot to the target.

Case 2: Partner pays attention to multiple targets

The robot's body orientation follows the vector from which the robot can see all of the targets without turning. Each angle between each pair of vectors from the robot to each target is calculated (Figure 4.13 left). *The robot's body orientation follows the median vector between two vectors that make the largest angle in all of the combination* (Figure 4.13, right).

4.3.4.3 Constraint of Partner's Field of View

Our previous research showed that constraint of a partner's field of view is important and that both the presenter robot and the object need to be in the listener's field of view. The results of our current observation also support this constraint. Thus, we defined the constraint of partner's field of view as below.

- *The angle between partner's body orientation and the vector from partner to robot must not be over 90 degrees* (Figure 4.14).

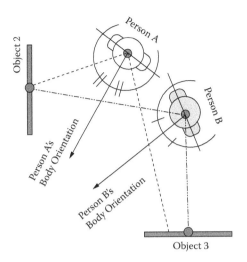

FIGURE 4.12
Spatial arrangement when people pay attention to multiple targets.

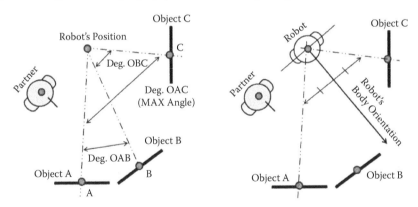

FIGURE 4.13
Body orientation.

4.3.4.4 Constraint of Robot's Field of View

Our previous research also showed that constraint of the robot's field of view is important and that the presenter robot needs to look both at its partner and the object. The results of our current observation also support this constraint. Thus, we defined the constraint of robot's field of view as shown below.

- *The angle between robot's body orientation and the vector from robot to partner should not be over 90 degrees* (Figure 4.15a).
- *All angles between robot's body orientation and each vector from robot to each object must not be over 75 degrees* (Figure 4.15b).

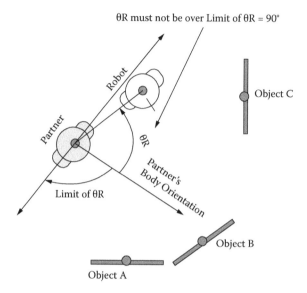

FIGURE 4.14
Constraint of partner's field of view: (a) toward partner, (b) toward each object.

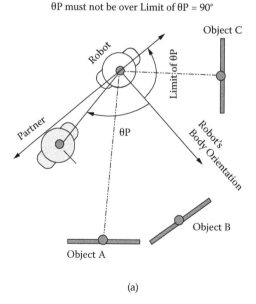

(a)

FIGURE 4.15
Constraint of robot's field of view.

Each θO must not be over Limit of θO = 75°

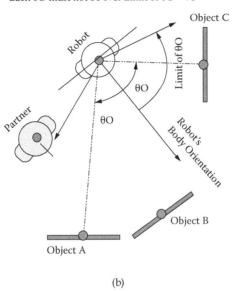

(b)

FIGURE 4.15 (continued)

4.3.5 Position Model for a Presenter Robot

To summarize the observation analysis, we established the following model for the robot to appropriately control its position:

$Value\_Of\_Place(Px) =$

$$\left(\left(PROX\_P(Px) + \sum PROX\_O(Px)\right) \times FOV\_P(Px) \times FOV\_R(Px)\right) \Big/ Dist$$

$$PROX\_P(Px) = \begin{cases} 1 \ (1100 \, mm < Dist. \ between \ Px \ and \ Partner < 1300 \, mm) \\ 0 \ (otherwise) \end{cases}$$

$$PROX\_O(Px) = \begin{cases} 1 \ (1000 \, mm < Dist. \ between \ Px \ and \ each \ Object < 1200 \, mm) \\ 0 \ (otherwise) \end{cases}$$

$$FOV\_P(Px) = \begin{cases} 1 \ (Partner's \ field \ of \ view < 90°) \\ 0 \ (otherwise) \end{cases}$$

$$FOV\_R(Px) = \begin{cases} 1 \ (\textit{Robot's field of view toward partner} < 90° \\ \qquad \&\& \quad \textit{Robot's field of view towatrd each object} < 75°) \\ 0 \ (\textit{otherwise}) \end{cases}$$

Dist = Dist.between Px and Robot current position

Here, *Px* is a possible position for a presenter. *PROX_L(Px)* is a function that indicates the constraint of proximity to listener. *PROX_O(Px)* is a function that indicates the constraint of proximity to an object. *FOV_L(Px)* is a function that indicates the constraint of the listener's field of view. *FOV_O(Px)* is a function that indicates the constraint of the presenter's field of view.

Position *Px* with maximum value *Value(Px)* must be chosen as the optimal standing position.

4.3.6 Implementation

The model was implemented in a humanoid robot, Robovie. As the purpose is to test effectiveness of the model, the study was conducted with a Wizard of Oz (WoZ) method to supplement the robot's lack of capability to autonomously address some issues that its software does not consistently address by itself. Also, a motion-capture system was used to stably recognize people's position and body orientations.

4.3.6.1 System Configuration

Figure 4.16 shows the system configuration that consists of a humanoid robot, a motion-capturing system, and a robot controller (software). We used the

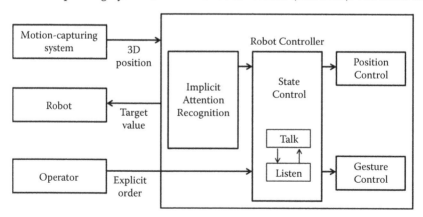

FIGURE 4.16
System configuration. Single-target setting; comparison setting.

humanoid robot named "Robovie" that is characterized by its human-like body expressions [18]. Its body consists of eyes, a head, and arms that generate complex body movements required for communication. In addition, it has two wheels for movement. Markers are attached to its head and arms. The motion-capturing system acquires body motions and outputs the position data of markers to the robot controller. Based on this data, the robot controller plans the robot's behavior.

4.3.6.2 Robot Controller

Figure 4.16 also shows an outline of the robot controller that consists of five units: *implicit attention recognition, state control, position control,* and *gesture control.* We are interested in the state where the robot automatically detects partner's attention and stands at the optimal position according to his/her attention state. On the other hand, it is difficult for the current robot to consistently hear what the listener is asking. Thus, we adopted a semi-autonomous system in the next section. When the listener asks the robot to present an object, the operator selected an object to be presented by the robot and input the target object to the robot controller. Based on the instructions, the robot controller automatically performed the robot's standing position and gestures.

4.3.6.3.1 Implicit Attention Recognition

This unit recognizes the listener's attention from implicit cues. The mechanism of recognition follows to Section 3.

4.3.6.3.2 State Control

This unit controls the state, which is either Talk or Listen, based on the situation. When the operator teaches the target object to the robot controller, the state changes from Listen to Talk. When the operator gives a Listen command to the robot controller, the state changes from Present to Listen.

4.3.6.3.3 Position Control

When the implicit attention recognition unit detects the partner's attention, this unit controls the robot's standing position based on our proposed model. The search area seeks the optimal standing position for the presenter. A grid as a possible 10 × 10 cm standing position divides the search area. This module estimates the values all of the grids in the search area and selects the one with the highest value as the optimal standing position.

4.3.6.3.4 Gesture Controller

The gesture controller manipulates the robot based on the state and the results of the position controller. When the state is Talk, this controller makes the robot maintain eye contact with the listener and points at the object.

4.3.7 Evaluation Experiment

We conducted an evaluation experiment to confirm the effectiveness of our proposed model.

4.3.7.1 Conditions

To verify the effectiveness of our proposed model, we created the following three conditions:

- *Stop condition*: The robot only responds to user's explicit attention. As soon as the robot received the partner's explicit order, the robot begins to turn and explain about object.
- *Explicit attention-shift condition*: The robot only respond to user's explicit attention. As soon as the robot received partner's explicit order, the robot begins to approach at the optimal position based on our proposed position model. After arriving at the optimal position, the robot begins to explain about object.
- *Implicit attention-shift condition*: The robot can respond to user's both explicit and implicit cues for attention shifting. Before the robot received its partner's explicit order, the robot finishes approaching the optimal position based on our proposed position model. As soon as the robot receives its partner's explicit order, the robot begins to explain the object.

The experiment had a within-subject design, and the order of all experimental trials was counterbalanced. Each participant performed in all three conditions.

4.3.7.2 Procedure

A total of 17 paid undergraduate students (10 men and 7 women whose age averaged 21) participated in this experiment. None of them majored in robotics.

We created a situation in which the robot answered participants' questions about their attention target(s). To verify the effectiveness of the model in various situations, we prepared two settings. In the first setting, participants paid attention to only one target (Figure 4.17a). In the second setting, participants paid attention to two targets (Figure 4.17b). Participants experienced both settings.

Single target setting: Participants moved in front of one panel. After arriving at the position, they read the panel and called the robot. As soon as participants called the robot, it turned to the participant, established eye contact, and said, "Yes, may I help you?" Participant then asked the robot, "What do you think about this phone?" while pointing toward it. The robot said, "O.K." and began to move toward the target position based on each condition. After

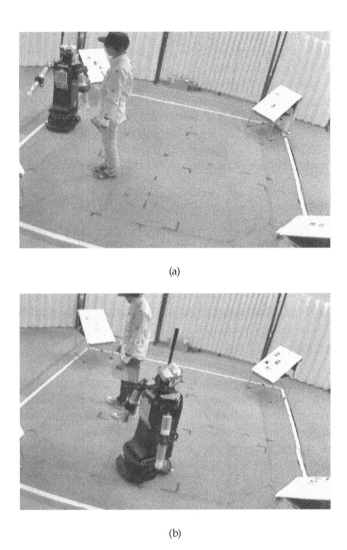

(a)

(b)

FIGURE 4.17
Experimental scene.

arriving at the target position, the robot pointed toward the phone and gave a brief opinion; for example, "This is a waterproof phone. It's good for you if you want to use in a bathroom." Participants evaluated the robot's behavior by answering a short questionnaire after each trial. Following this procedure, participants requested and received information on four different panels.

Comparison setting: Participants moved around the center of the experimental area to compare two different panels. After arriving at the position, they thought about which phone is expensive for a while. As soon as participants called the robot, it turned to the participant, established eye contact, and said, "Yes, may I help you?" Participants then asked the robot, "This phone and this phone, which phone is expensive?" while pointing toward each panel. As soon as the robot received the partner's request, it said "OK," and began to move toward the target position based on each condition. After arriving at the position, the robot answered participants' question; for example, "The price of this phone is 35,000 yen. This phone is 50,000 yen. So, this phone is more expensive" while pointing to each panel. The partner evaluated the robot's behavior by answering a short questionnaire. In this way, the partner requested and received information on four different pairs of panels.

4.3.7.3 Measures

We obtained participants' subjective impressions with a questionnaire. Following questions were included on 1-to-7 point Likert scales.

- **Place**. How comfortable/uncomfortable did you feel about the position of the robot when it talked with you?
- **Timing**. How comfortable/uncomfortable did you feel about the timing when the robot began explaining its opinion?
- **Attentiveness**. How attentive/inattentive was the robot?

The "Place" question measured how comfortable participants were with the place where the robot stood, based on our proposed model formed by observations of human communication. The "Timing" question measured their comfort with the time it took the robot to respond during interaction. We created the "Timing" question to measure how well our proposed model estimated the place where the robot should stand and talk. The "Attentiveness" question measured how much they liked the robot during interaction. The ratings of each item were averaged in four trials per each setting.

4.3.7.4 Hypothesis and Predictions

Since we developed a spatial position model for a robot to stand as a comfortable position during conversations, we expect that with this model the robot would provide a better impression on its position.

Since the attention-shift model is developed for enabling the robot to engage in a task with the appropriate timing, we expect that the robot with the attention-shift model would provide a better impression on its timing.

Since the goal of recognizing both implicit and explicit cues is to provide a robot with the capability to maintain an attentive interaction, we expect that the robot with both capabilities will better than the one with the recognition of explicit cue, and the one with the recognition of explicit cue will be better than the one without it.

Based on these considerations, we made the following predictions:

Prediction 1: Participants will evaluate the place of the robot better in *explicit attention-shift* and *implicit attention-shift* conditions than in *stop* condition.

Prediction 2: Participants will evaluate the timing better that the robot performs better in *implicit attention-shift condition* than in *stop* and *explicit attention-shift condition*.

Prediction 3: Participants will evaluate the attentiveness in *implicit attention-shift conditions* as the best, and then evaluate *explicit attention-shift condition* better than in *stop condition*.

4.3.8 Results

4.3.8.1 Verification of Prediction 1

Figure 4.18 shows the results from the questionnaire on *place*. We conducted a repeated-measures analysis of variance (ANOVA) for the *single target* setting

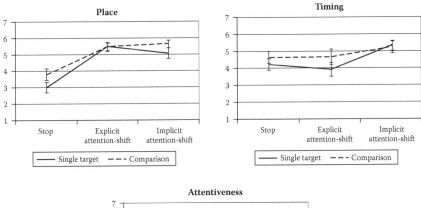

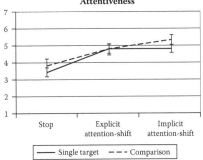

FIGURE 4.18
Experimental results.

that showed a significant difference between conditions (F[2,32] = 40.55, $p < .01$). We also conducted multiple comparisons with Bonferroni methods that showed that the *stop* condition was less preferred than *explicit attention-shift* condition ($p < .01$) and *implicit attention-shift* conditions ($p < .01$). Difference between *explicit attention-shift* and *implicit attention-shift* condition was also significant ($p = .05$).

We also conducted a repeated-measures ANOVA for *comparison* setting that showed a significant difference between conditions (F[2,32] = 31.64, $p < .01$). We also conducted multiple comparisons with Bonferroni methods that showed that *stop* condition was less preferred than *explicit attention-shift* condition ($p < .001$) and *implicit attention-shift* conditions ($p < .001$). These results confirmed our first prediction.

4.3.8.2 Verification of Prediction 2

Figure 4.18 shows the results from the questionnaire on *timing*. We conducted a repeated-measures analysis of variance (ANOVA) for *single target* setting that showed a significant difference between conditions (F[2,32] = 14.22, $p < .01$). We also conducted multiple comparisons with Bonferroni methods that showed that *implicit attention-shift* condition was preferred over *stop* condition ($p < .05$) and *explicit attention-shift* conditions ($p < .001$).

We also conducted repeated-measures ANOVA for *comparison* setting that showed no significant difference between conditions (F(2,32) = 2.117, $p = .137$).

Thus, prediction 2 was confirmed for the single target settings, but not confirmed for the comparison settings.

4.3.8.3 Verification of Prediction 3

Figure 4.18 shows the results from the questionnaire on *attentiveness*. We conducted a repeated-measures analysis of variance (ANOVA) for *single target* setting that showed a significant difference between conditions (F[2,32] = 16.54, $p < .01$). We conducted multiple comparisons with Bonferroni methods that showed that *stop* condition was less preferred than the *explicit attention-shift* condition ($p < .05$) and *implicit attention-shift* condition ($p < .05$). There were no significant differences between the *explicit attention-shift* and *implicit attention-shift* conditions ($p = 1.00$).

We also conducted repeated-measures ANOVA for the *comparison* setting that showed a significant difference between conditions (F[2,32] = 10.12, $p < .01$). We conducted multiple comparisons with Bonferroni methods that showed that *stop* condition was less preferred than *explicit attention-shift* condition ($p < .005$) and the *implicit attention-shift* condition ($p < .005$). There were no significant differences between the *explicit attention-shift* and *implicit attention-shift* conditions ($p = .160$).

Thus, prediction 3 was partially confirmed, revealing that the *implicit attention-shift* and *explicit attention-shift* conditions are better than the *stop* condition, but the predicted difference between the *implicit attention-shift* and *explicit attention-shift* condition are not supported.

4.3.9 Limitations

Since our model is built as a proof-of-concept, there are many limitations in its current form. One of the notable limitations is that the objects that people focus their attention on are limited to big objects; if there are small objects, we should combine other modalities, such as gaze. We believe that this is a reasonable focus; the mobile robot would engage in interaction with such big objects in many situations in the future, such as a museum-guide robot that talks about exhibits, and a shopkeeper robot that explains and helps people to compare products, such as computers and home appliances.

4.3.10 Conclusions

This chapter reports a study for modeling natural behavior in a situation where a robot shares attention implicitly with people. It involves two models: *implicit cue for attention-shifting* and *spatial position*. The analysis shows that people's body position and orientation work as implicit cues for sharing attention, which turned into three submethods: *estimate attention*, *detect attention-shift*, and *predicting attention during attention shifting*. The spatial position model is developed based on our previous model [1] with an extension for multiple objects.

The experimental result mostly reveals the effectiveness of the model, but there is no significant difference in timing for the situations when they compared multiple objects. We consider it due to the hardware limitation. While humans' attention shift is only within a few seconds, the robot needs at least 10 seconds for the process, which is largely due to the slowness of its motor.

References

1. Yamaoka, F., Kanda, T., Ishiguro, H., and Hagita, N. (2008). How close? A model of proximity control for information-presenting robots, *ACM/IEEE 3rd Annual Conference on Human-Robot Interaction*, pp. 137–144.
2. Moore, C. and Dunham, P. J. (Ed.). (1995). Joint attention: Its origins and role in development. Lawrence Erlbaum Associates.
3. Kozima, H. and Yano, H. (2001). A robot that learns to communicate with human caregivers. In Proceedings of the First International Workshop on Epigenetic Robotics.
4. Nagai, Y., Hosoda, K., Morita, A., and Asada, M. (2003). A constructive model for the development of joint attention. Connection Science, No. 4, pp. 211–229, 74, 2000.

5. Scassellati, B. (2000). Theory of mind for a humanoid robot. In Proc. of the 1st IEEE/RSJ International Conference on Humanoid Robots, pp. CD–ROM.

6. Mutlu, B., Yamaoka, F., Kanda, T., Ishiguro, H., and Hagita, N. (2009). *Nonverbal Leakage in Robots: Communication of Intentions through Seemingly Unintentional Behavior.* Paper presented at the ACM/IEEE Int. Conf. on Human-Robot Interaction (HRI2009).

7. McNeill, D. (1987). *Psycholinguistics: A New Approach.*

8. Sugiyama, O., Kanda, T., Imai, M., Ishiguro, H., and Hagita, N. (2006). Humanlike conversation with gestures and verbal cues based on a three-layer attention-drawing model, *Connection Science*, 18(4), pp. 379–402.

9. Breazeal, C., Kidd, C. D., Thomaz, A. L., Hoffman, G., Berlin, M. (2005). *IEEE/RSJ International Conference on Intelligent Robots and Systems*, pp. 383–388.

10. Kendon, A. (1990). *Conducting Interaction–Patterns of Behavior in Focused Encounters,* Cambridge University Press.

11. Dautenhahn, K., Walters, M. L., Woods, S., Koay, K. L., Nehaniv, C. L., Sisbot, E. A., et al. (2006). How may i serve you? A robot companion approaching a seated person in a helping context. Paper presented at the ACM/IEEE Int. Conf. on Human-Robot Interaction (HRI2006).

12. Hüttenrauch, H., Eklundh, K. S., Green, A., and Topp, E. A. (2006). Investigating spatial relationships in human-robot interactions. Paper presented at the IEEE/RSJ Int. Conf. on Intelligent Robots and Systems (IROS2006).

13. Sisbot, E. A., Alami, R., Simeon, T., Dautenhahn, K., Walters, M., Woods, S., Koay, K. L., and Nehaniv, C. (2005). Navigation in the presence of humans, *IEEE-RAS International Conference on Humanoid Robots.*

14. Pacchierotti, E., Christensen, H. I., and Jensfelt, P. (2006). Design of an office guide robot for social interaction studies, In *Proc. Int. Conf. Intelligent Robots and Systems (IROS2006)*, pp. 4965–4970.

15. Gockley, R., Forlizzi, J., Simmons, R. (2007). Natural person-following behavior for social robots, *In Proc. Int. Conf. on Human-robot interaction (HRI2007)*, pp. 17–24.

16. Kendon, A. (1990). *Conducting Interaction: Patterns of Behavior in Focused Encounters*: Cambridge University Press.

17. Moore, C., and Dunham, P. J. (1995). *Joint Attention: Its Origins and Role in Development*: Lawrence Erlbaum Associates.

18. Kanda, T., Ishiguro, H., Imai, M., and Ono, T. (2004). Development and evaluation of interactive humanoid robots, *Proceedings of the IEEE*, Vol. 92, No. 11, pp. 1839–1850.

4.4 A Model for Natural and Comprehensive Direction Giving

Yusuke Okuno, Takayuki Kanda, Michita Imai,
Hiroshi Ishiguro, and Norihiro Hagita

ATR Intelligent Robotics and Communication Laboratory

4.4.1 Introduction*

In Chapter ??, we introduced our field trials. In a shopping mall, we made the robot provide directions. We found such direction given by a robot useful. A robot has a number of appropriate features for direction giving; since it is physically co-located with people, it can proactively approach a person who needs such information, and then provide it "naturally" with its human-like body properties. While what was used in the field trial were simple directions, we are better prepared to understand now what good direction giving involves.

What constitutes good direction giving from a robot? If the destination is within a visible distance, the answer might be intuitive. A robot would say "The shop is over there" and point. However, since the destination is often not visible, a robot needs to utter several sentences. Moreover, it would be expected to be accompanied with gestures. We designed our robot's behavior to enable the listener to intuitively understand the information provided by the robot. This chapter illustrates how we integrate three important factors—*utterances*, *gestures*, and *timing*—so that the robot can conduct appropriate direction giving.

4.4.2 Related Works

4.4.2.1 Utterance

With respect to humans' direction giving, Daniel et al. makes the following insightful observations: "A remarkable fact about direction giving is that they do not always make it easy for people to reach their goal." They add that "ambiguous and confusing descriptions are known to be inefficient," and "descriptions that are too long and too detailed, however correct, become too difficult to memorize." They also established a "skeletal description," which consists of a series of sentences where each sentence contains a pair that consists of a landmark and an action [1]. We follow their study to optimize a robot's verbal information.

Also in robotics, a robot gives directions [2]; but this previous study did not include landmarks in verbal information, since it was conducted in a simple corridor setting that did not require landmarks for direction giving.

4.4.2.2 Gesture

In human communication studies, a couple of studies have revealed humans' gesture in direction giving. Allen found that people performed deictic gesture (68%) more often than other gestures such as iconics (20%), beats (11%), and metaphorics (2%). This study also found that people who spoke fast do

---

* This chapter is a modified version of a previously published paper Okuno, Y., Kanda, T., Imai, M., Ishiguro, H., and Hagita, N. (2009). Providing route directions: Design of robot's utterance, gesture, and timing, *ACM/IEEE Int. Conf. on Human-Robot Interaction (HRI2009)*, pp. 53–60.

more gesturing, which implies that gestures make utterances fluid [3]. In a study of the natural behavior of an embodied conversation agent, human behavior in direction giving was investigated. People usually embrace the route perspective (the perspective of the person following the route) (54.1%) rather than a survey perspective (16.3%) [4,5].

Kita found that when a person utters a directional expression such as "Turn right," he performs representational gestures to describe the action/direction of "right." The explaining person turns his torso orientation so that his "right" side matches with the direction of the route that will "turn right" [6]. In other words, this finding suggests that it is natural for a human speaker to coordinate his/her body orientation when describing a direction.

One previous study dealt with a robot's direction giving. Ono et al. revealed that appropriate body orientation is important for conveying such directions as "right" and "left." They also suggested a possibility that gaze and body movements of the robot would reduce the time required for subjects to reach their destination [2]. Our study aims to move one step beyond Ono's study to reveal how gestures contribute to the direction giving.

4.4.2.3 Timing

In this chapter, we explore an aspect of pause duration between the robot's utterances. This pause resembles the one in the following passage: "Please go straight. (*pause*) Then you will see a post office." Unlike the switching pause in turn-taking [7], this pause should not cause the switch of speakers. It is known that pauses between sentences are highly variable and depend on speakers and their tasks, often ranging from 300 to 1000 ms [8,9]. Pause durations also depend on cultures. For instance, Campione et al. revealed that the average duration varies from 400 to 550 ms among five languages [10]. The pause duration might be slightly shorter in Japanese, as Nagaoka et al. found by analyzing a large corpus of spoken Japanese. They reported that the majority of pauses between sentences of the same speaker are less than 400 ms [11]. Although these previous studies modeled the pause duration between sentences by modeling human speech, in this study we focus on modeling from the listener's perspective; to our knowledge, this is yet an underexplored aspect.

4.4.3 Modeling Robot's Direction Giving

We modeled a process to generate a robot's direction giving and divided it into three models: *utterances*, *gestures*, and *timing*. Figure 4.19 shows the information flow among these three models. First, a robot generates *utterance*. Second, it combines an *utterance* with a *gesture*. Finally, it expresses direction giving with appropriate *timing*. For each model, we took different approaches, as summarized in Table 4.1.

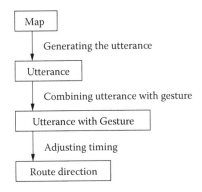

FIGURE 4.19
A process to generate robot's direction giving.

TABLE 4.1

Modeling of direction giving

Model	Design Method
Utterance	Literature review
Gesture	Design consideration
Timing	Modeling from human behavior

4.4.3.1 Utterance

Direction giving is described as a process of providing knowledge about directions by means of utterances and gestures in combination. Usually, the source of the information is a geographical map (e.g., Figure 4.20) that contains locations of buildings and streets. A robot decides the way to the destination from the map, and then makes sentences to describe the way. For developing the utterance model, we did a literature review and decided to rely on the work conducted by Daniel et al. [1], who concluded that route description should contain minimal information that is neither too short to avoid ambiguity nor too detailed. From this standpoint, they proposed a "skeletal description" that contains minimal sets of information.

A "skeletal description" [1] consists of a series of sentences, and each sentence consists of a pair comprising a landmark and an action. An action is an instruction about walking behavior, such as "go straight," "turn left," or "turn right." A landmark is an easy-to-find building in the environment where people are instructed to take an action, such as a bank, a post office, or a library. Thus, a sentence should be something like, "Turn left at a bank."

Following the "skeletal description," the robot uses these sentences to provide information about how to reach the destination. Table 4.2 shows an example of utterances for direction giving between two places on the map shown in Figure 4.20. In the table, *S1-6* indicates sentences uttered by the robot, and *P1-5* indicates pauses between them.

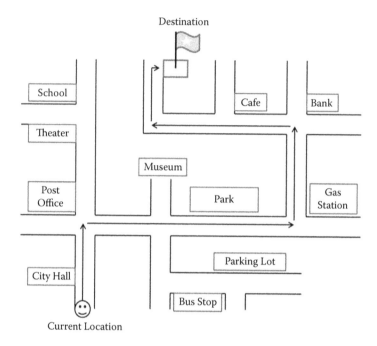

FIGURE 4.20
Street map for locating directions to objective.

TABLE 4.2

Utterance based on skeletal description

S1	<u>Go straight</u> this way.
	(Action 1)
P1	(Pause 1)
S2	<u>Turn right</u> at a <u>post office</u>.
	(Action 2) (Landmark 2)
P2	(Pause 2)
S3	<u>Turn left</u> at a <u>gas station</u>.
	(Action 3) (Landmark 3)
P3	(Pause 3)
S4	<u>Turn left</u> at a <u>bank</u>.
	(Action 4) (Landmark 4)
P4	(Pause 4)
S5	<u>Turn right</u> at a <u>library</u>.
	(Action 5) (Landmark 5)
P5	(Pause 5)
S6	Then you will reach the destination.

TABLE 4.3

Four types of gestures [2,3,4,5,6,12]

	Deictic	Orienting Body Direction	Iconic (Landmarks)	Beat
Useful for speakers	Yes	Yes	Yes	Yes
Useful for listeners	Definitely yes	Yes in the beginning unclear for later	Maybe	No (unless utterance is redundant)
Robots can express well	Yes	Yes	Difficult	Maybe

4.4.3.2 Gesture

In direction giving, since people also use gestures (see Section 4.4.2.2), we wonder what kind of robot gestures could help people understand direction giving. From the literature [2,3,4,5,6,12], we retrieved gestures often used in direction giving and made a list that contains the following four types: *deictic gesture, orienting body direction, iconic gesture (expressing landmarks)*, and *beat gesture*. Table 4.3 shows a summary of these gesture types. The following three aspects are considered.

4.4.3.2.1 Aspect 1: Is the Gesture Accompanied by Speech?

Gestures are produced by speakers in facilitating their speech [13]. All of the listed gestures are produced in the process of formulating speech, because they are used in the context of direction giving.

4.4.3.2.2 Aspect 2: Does the gesture help human listeners understand?

Gestures help listeners understand utterances when a message is complicated or unclear (e.g., [14,15]). Since our utterance contains sufficient information, the information that could be conveyed by gesture is also conveyed by utterances. Thus, gesture could be considered as a redundant message in our case (for example, pointing the "right" direction and saying "Turn right").

Under this assumption, we considered that "deictic gestures" would be useful, as pointing out the direction could visually provide a clear supplement for such an utterance as "Turn right" with its absolute direction to walk. This is the gesture that is most often used in direction giving [3]. At least, there is little risk that it would cause misunderstanding.

Orienting body direction would be useful in cases where "right" and "left" are inconsistent between the speaker and the listener, e.g., when facing each other [2]; however, once they establish a direction (e.g., standing to the side) changing body orientation does not help much. On the other hand, one risk of using "orienting body direction" is that when the robot frequently orients its body orientation, if people do not follow, they lose the coordination of body orientation; thus, it is not clear whether this would be useful.

Iconic gestures (expressing landmarks) could be useful, though they depend on the available landmarks. Some landmarks are easily represented as icons, but many are difficult, such as gas stations, banks, or libraries. There could be a risk here; if people do not understand the iconic gesture of the robot, they might be confused about it.

Beat gestures are usually used for emphasizing important parts in speech; however, since "skeletal description" contains minimal information, they are not so useful in our case. Also, there could be a risk that people would not understand the meaning of the beat gesture of the robot.

4.4.3.2.3 Aspect 3: Can Robots Express the Appropriate Gesture?

This aspect depends on the robot's shape. Robots have often limited degrees of freedom in comparison to humans. In our case, the robot can perform deictic and beat gestures, and orient its body direction, but it had difficulty performing iconic gestures well because they usually require hands and fingers to create shapes, which was impossible for our robot.

Overall, we decided to use deictic gestures after the robot and listening person orient their body directions. We did not use iconic and beat gestures because it was unclear whether it would work positively or negatively; we intended to reveal the usefulness of the most promising gesture, deictic gesture, and placed the other gestures for a topic for the future study.

Figure 4.21 shows the deictic gesture we designed. The deictic gesture is used to point the absolute direction. (It is the iconic gesture that is used to represent the action of the movements.) For example, for the case when the robot utters the following consecutive sentences: "Go straight this way" (Figure 4.21a). "Turn right at a post office" (Figure 4.21b). "Turn left at a gas station" (Figure 4.21a). Since the deictic gesture points at the absolute direction, it performs gesture (a) but not the gesture (c) for the last phrase "Turn left."

4.4.3.3 Timing

The robot pauses between sentences when it speaks (Figure 4.22). This model of *timing* decides the duration of these pauses. We take two different approaches for modeling the timing: from the speaker's perspective (i.e., the natural timing for speaking) and from the listener's perspective (i.e., the time needed to understand). Later, we experimentally decided which one enables a robot to provide better direction giving. (The experiment is explained in Section 4.4.4.)

4.4.3.3.1 Modeling from Speaker's Timing

Modeling of human speaker behavior is commonly used for creating naturalness in computers, for example, speech synthesis or CG agent [4,8,9,11,16,17,18,19,20,21]. With this modeling approach, the robot behaves similarly to humans. Thus, a possible hypothesis is that a robot could naturally perform direction giving based on the timing modeled from a human speaker's timing.

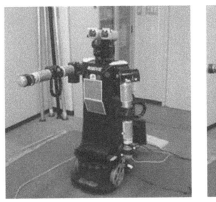
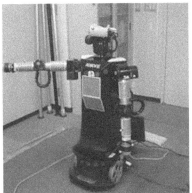

(a)   (b)

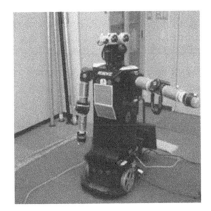

(c)

FIGURE 4.21
Deictic gestures.

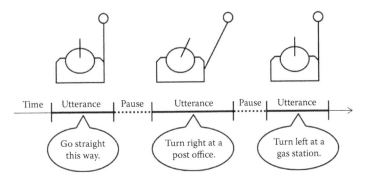

FIGURE 4.22
Pauses during route guidance.

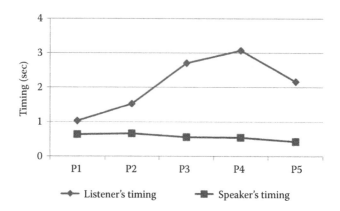

FIGURE 4.23
Timing model.

To model the timing of speakers, we measured their pause durations. We asked six students in our laboratory who did not know the purpose of the study to read the sentences generated with the "skeletal description" described in Section 3.1. The description consisted of six sentences. They looked at the description for a minute, and then read it naturally to a listener. The listener was standing in front of the speaker, but did not provide a particular reaction (e.g., nodding) to the speaker. We measured the pause duration between sentences.

Figure 4.23 shows the average duration of the pauses of the six speakers. Pi represents a pause after i-th sentence (the same symbols used in Table 4.2). The *P1* to *P5* average ranged between 0.42 and 0.66 seconds, which seems a reasonable value in comparison with previous speech synthesis studies [8,9,11].

Based on this "speaker" model, the robot used these measured *P1* to *P5* for the pause duration in direction giving.

4.4.3.3.2 *Modeling from Listener's Timing*

An alternative timing approach is to model the time that people take to understand the utterances. Listening to direction giving is a cognitively demanding task that requires the comprehension of spatial relationships and the memorization of routes and landmarks. Thus, making a robot who takes enough time after each sentence is reasonable and could increase understanding in listeners.

We modeled the time that people take to understand direction giving. For the data collection, a robot performed direction giving to a human participant. We used the same robot as in the experiment reported in Section 4. The robot spoke the sentences generated based on the "skeletal description" described in Section 3.1. The description consists of six sentences. In the data collection, the robot uttered sentences one by one; participants were asked to verify that they understood the utterance after each sentence the robot uttered; after verification, the robot started the next sentence. In this data collection, the robot was controlled by the Wizard-of-Oz method.

Eight university students participated in this experiment. We measured the time between the end of the robot's utterance and when participants reported that they understood. Each participant repeated this measurement four times for different routes. Figure 4.23 (upper line) shows the duration average required by listeners to understand the utterance. It is the average of eight participants with four trials. On average, P1 to P5 ranged between 1.03 and 3.07 seconds. Based on this "listener" model, the robot uses these measured P1 to P5 for pause durations in the direction giving. These values are quite long in comparison to those found in previous speech synthesis studies [8,9,11].

4.4.4 Evaluation Experiment

We conducted an evaluation experiment to identify the effect of gestures and to find a better model of timings.

4.4.4.1 Hypothesis and Predictions

Since previous studies indicated the usefulness of human gestures for human listeners, we thought that the robot's gestures would help listeners understand direction giving, even when the gesture conveyed a redundant message represented in parallel by utterances, such as pointing "right" direction while saying "Turn right."

Since Figure 4.23 shows that listeners need more time to understand direction giving than the pause duration of speakers, we thought that a speaker-based model would not offer enough time for listeners to understand.

Based on these considerations, we made the following predictions:

1. When participants listen to the direction giving with gestures, the correctness scores and easiness ratings will outperform the cases when they listen without gesture.
2. When participants listen to the direction giving with the timing of the listener-based model, correctness scores and easiness ratings will outperform the cases when they listen to the direction giving with the timing of the speaker-based model.

4.4.4.2 Method

4.4.4.2.1 Participants

Participating in our paid experiment were 21 undergraduate, native Japanese speakers (14 males and 7 females).

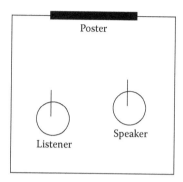

FIGURE 4.24
Experiment environment.

4.4.4.2.2 *Settings*

We used "Robovie," a 1.2 m tall communication robot with a 0.5 m radius whose human-like upper body is designed for communication with humans. It has a head (3 DOFs), eyes, and arms (4*2 DOF). A speaker was attached on its head. With its 4-DOF arms, it can perform deictic gestures (Figure 4.21).

Figure 4.24 shows the experimentation environment, which is a 3 × 3 m space separated from the rest of the room by partitions. An A0 size picture of a way in a town is presented on the face of one partition. A speaker (i.e., a human or a robot depending on condition) and a listener (participant) stood near the picture.

4.4.4.2.3 *Conditions*

There were two conditions for both gestures and timing. The robot gave directions based on the description shown in Section 4.4.3.1. They were based on the "skeletal description," and each had six sentences. (Scene of the experiment is also available as a video attachment.)

4.4.4.2.3.1 *Gesture*

With: The robot performed the deictic gestures described in Section 4.4.3.2 (Figure 4.21).

Without: The robot did not do any gestures. Thus, no arm movements were expressed.

4.4.4.2.3.2 *Timing*

Speaker: The robot uttered sentences based on the speaker-based model described in Section 4.4.3.1.

Listener: The robot uttered sentences based on the listener-based model described in Section 4.4.3.2.

4.4.4.2.4 Procedure

The experiment was a within-subject design. Participants repeated the session for all conditions. The order of sessions was counterbalanced. The given route and the destination were different every time. They received a 10 minute break between each session.

At the first session, participants were instructed to imagine getting lost at an unfamiliar place on their way to a famous restaurant. They were instructed to ask about a route, to learn it, and then to draw it on a piece of paper after listening to the direction giving. They were also told that they had to learn both landmarks and actions to reach the destination, since the map was not on a grid. Participants were positioned at the "listener" position in Figure 4.24, and the robot/human speaker stood at the "speaker" position.

At each session, the participants listened to the direction giving. After the direction giving had been provided, they drew a map and completed questionnaires. Note that we prohibited participants from asking for the direction giving again; thus, the direction giving were given only once per condition.

4.4.4.2.5 Measurement

We conducted two types of evaluations. One was about the task performance; that is, how well participants understood the direction giving.

Correctness: We asked participants to draw a map of the route that was explained during the experiment. We counted the number of correct actions and landmarks on the maps. Each description had four landmarks and four actions; we excluded the first sentence from the score (S1 in Table 4.2) because it was always "Go straight on this street" and all participants answered it correctly. The score is ranged from 0 to 8. For example, three correct landmark and two correct actions were scored as 5.

The participants were also asked to provide comments about their impressions with respect to the following items on a 1–7 scale where 1 stands for the lowest evaluation and 7 for the highest.

Easiness: How easy/difficult was it to understand the route?

Naturalness: How natural/unnatural were the direction giving?

4.4.5 Results

4.4.5.1 Verification of Predictions

For the correctness score results (Figure 4.25), a two-way repeated-measures analysis of variance (ANOVA) was conducted with two within-subject factors, *gesture* and *timing*. A significant main effect was revealed in both the gesture factor ($F(1,20) = 16.055$, $p < .005$) and the timing factor ($F(1,20) = 6.757$, $p < .05$), but no significance was found in the interaction within these factors ($F(1,20) = .323$, $p = .576$).

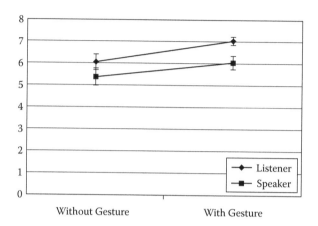

FIGURE 4.25
Experimental results of correctness scores.

For the results of the easiness ratings (Figure 4.26a), a two-way repeated-measures ANOVA with two within-subject factors, *gesture* and *timing*, was conducted. A significant main effect was revealed in both the gesture factor ($F(1,20) = 13.945$, $p < .005$) and the timing factor ($F(1,20) = 12.105$, $p < .005$). Interaction within these factors was also significant ($F(1,20) = 6.448$, $p < .05$).

This interaction suggests that each of factors contributed to the easiness but the combination of two factors (with-gesture and listener-based model) did not improve the easiness twice. There was a simple main effect in the gestures in the speaker-based model ($F(1,20) = 28.027$, $p < .001$), while no significant main effect was observed in the gestures in the listener-based model ($F(1,20) = 2.077$, $p = .165$). There was a simple main effect in the timing in the without-gesture condition ($F(1,20) = 30.941$, $p < .001$), but no significant main effect was observed in the timing in the with-gesture condition ($F(1,20) = 1.709$, $p = .206$).

Overall, predictions 1 and 2 are supported.

4.4.5.2 Comparison of Naturalness Ratings

A two-way repeated-measures ANOVA was conducted with two within-subject factors, *gesture* and *timing*, for the result of naturalness (Figure 4.26b). A significant main effect was revealed in both the gesture factor ($F(1,20) = 8.928$, $p < .01$) and the timing factor ($F(1,20) = 12.308$, $p < .005$). Interaction within these factors was also significant ($F(1,20) = 5.525$, $p < .05$). We analyzed the simple main effects in the interaction within these factors. It was almost significant in the gestures in the listener-based model ($F(1,20) = 3.467$, $p = .077$), and significant in the gestures in the speaker-based model ($F(1,20) = 14.000$, $p < .005$), in timing in the with-gesture condition ($F(1,20) = 5.972$, $p < .05$) as well as the without-gesture condition ($F(1,20) = 15.838$, $p < .005$). Overall,

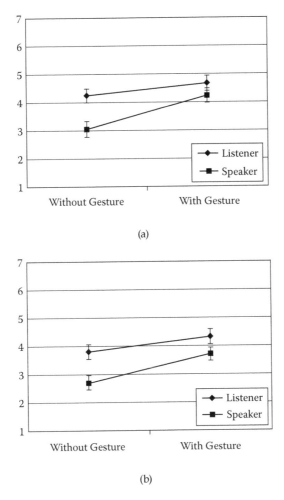

FIGURE 4.26
Experimental results of subjective evaluations: (a) easiness, (b) naturalness.

participants rated the naturalness ratings higher for the robot that uttered with gestures and with the timing of the listener-based model. The combination of the two factors did not improve the naturalness rating twice, as in the case of the easiness ratings.

4.4.6 Limitations

This study is one of the first systematic studies of robot's direction giving. The generality of the model is yet limited. For instance, the timing model depends on the number of sentences and complicity of contents. It also depends on the capability of speech synthesis. When we interviewed the participants, some of them mentioned the difficulty to listen to the robot's utterances due to its unusual accent and intonation. Technologies are advancing year by year, thus, in the future it would be possible for people to understand what the robot speaks in shorter time than what it was possible in this study. This aspect also depends on languages. That is, the finding is limited to Japanese.

4.4.7 Conclusion

This study reports the model for a robot's direction giving. Aiming to provide a comprehensive and natural direction giving, we learned from human communication studies. While utterance and gesture is modeled from such studies, we hypothetically built two models for timing, one learned from speakers, one from listeners. The 2 × 2 design experiment was conducted to test the effect of the gesture and timing submodel. The result revealed the effect of each submodel. Although the information from gesture is redundant to utterance in robot's direction giving, it was demonstrated that gesture made the direction giving more natural and comprehensive. That is, even though utterances provide enough information for a listener to understand the way, the gestures add information to the utterances. Moreover, the result confirmed that a listener-based timing model is more effective than a speaker-based timing model in terms of both naturalness and comprehensiveness.

   While previous studies for natural human–robot interaction often modeled human behavior in communication, the unique aspect of this study is that it modeled the listener's behavior, that is, a required time to comprehend speech and gestures that provides information. It is notable that such a listener-based model is not only demonstrated to be more comprehensive, but also perceived as more natural. We consider that such an understanding of human perspective would be a promising way to model complex natural interaction.

Acknowledgments

This research was supported by the Ministry of Internal Affairs and Communications of Japan.

References

1. Daniel, M., Tom, A., Manghi, E., and Denis, M. (2003). Testing the value of route directions through navigational performance. *Spatial Cognition and Computation.* 3, 4, 269–289.

2. Ono, T., Imai, M., and Ishiguro, H. (2001). A model of embodied communications with gestures between humans and robots. *Proc. Annual Meeting of the Cognitive Science Society (CogSci2001)*. 732–737.
3. Allen, G. (2003). Gestures accompanying verbal route directions: Do they point to a new avenue for examining spatial representations? *Spatial Cognition and Computation*. 3, 4, 259–268.
4. Kopp, S., Tepper, P.A., Ferriman, K., Striegnitz, K., and Cassell, J. (2008). Trading spaces: How humans and humanoids use speech and gesture to give directions. In Nishida, T. (Ed.), *Conversational Informatics: An Engineering Approach.*
5. Striegnitz, K., Tepper, P., Lovett, A., and Cassell, J. (2005). Knowledge representation for generating locating gestures in route directions. WS in *Spatial Language and Dialogue.*
6. Kita, S. (2003). Interplay of gaze, hand, torso orientation and language in pointing. In Kita, S. (Ed.), *Pointing: Where Language, Culture, and Cognition Meet.* Psychology Press, NY.
7. Jaffe, J., and Feldstein, S. (1970). *Rhythms of Dialogue.* Academic Press. New York.
8. Zvonik, E., and Cummins, F. (2002). Pause duration and variability in read texts. *Int. Conf. on Spoken Language Processing (ICSLP-2002)*. 1109–1112.
9. Zvonik, E., and Cummins, F. (2003). The effect of surrounding phrase lengths on pause duration. *European Conference on Speech Communication and Technology (EUROSPEECH-2003)*. 777–780.
10. Campione, E., et al. (2002). A large-scale multilingual study of silent pause duration. *Speech Prosody*. 199–202.
11. Komori, M., Yamamoto, Y., and Nagaoka, C. (2006). Manipulation of pause durations for the facilitation of understanding of speech played back at higher rate. *The Japanese Journal of Ergonomics*. 42, 2, 64–69 (in Japanese).
12. Kendon, A. (2004). *Gesture: Visible Action as Utterance.*
13. Krauss, R.M. (1998). Why do we gesture when we speak? *Current Directions in Psychological Science*. 7, 54–59.
14. Alibali, M. (2005). Gesture in spatial cognition: Expressing, communicating, and thinking about spatial information. *Spatial Cognition and Computation*. 5, 4, 307–331.
15. Trafton, J.G., et al. (2006). The relationship between spatial transformations and iconic gestures. *Spatial Cognition and Computation*. 6, 1, 1–29.
16. Mutlu, B., Hodgins, J.K., and Forlizzi, J. (2006). A storytelling robot: modeling and evaluation of human-like gaze behavior. *HUMANOIDS'06*. 518–523.
17. Nagaoka, C., et al. (2005). Influence of Response Latencies on Impression Evaluation of Speakers in Dialogues. Technical Report of IEICE. 104, 745, 57–60 (in Japanese).
18. Ogawa H., and Watanabe, T. (2001). InterRobot: speech-driven embodied interaction robot. *Advanced Robotics*. 15, 3, 371–377.
19. Yamamoto, M., and Watanabe, T. (2004). Timing control effects of utterance to communicative actions on embodied interaction with a robot. *IEEE Int. Workshop on Robot and Human Communication (ROMAN2004)*. 467–472.
20. Yamamoto, M., and Watanabe, T. 2006. Time lag effects of utterance to communicative actions on cg character-human greeting interaction. *ROMAN2006*. 629–634.
21. Zellner, B. 1994. Pauses and the temporal structure of speech. In Keller, E. (Ed.), *Fundamentals of Speech Synthesis and Speech Recognition*. 41–61.

5

Sensing Systems: Networked Robot Approach

5.1 Introduction

Sensing system is the fundamental components in most of robotic systems. Nowadays, we humans live in the world with ubiquitous sensors, and access to the Internet via smart phones and computers to refer to information in distant locations. In similar metaphor, as robots are networked, large part of sensing can be done with ubiquitous sensors in network robot system. The studies in the Chapter 2 already shows examples of network robots working in a real fields. In this chapter we specifically introduce a couple of techniques in details.

There are several sensors used for sensor networks. This book focuses on the following five sensors:

a. Floor sensors

b. Active tags

c. Passive tags

d. Laser scanners and cameras

e. Skin sensors

The floor sensor is a kind of pressure sensor that covers a wide area. It can detect human and robot footprints mainly in an indoor environment.

Both the active and passive tags are used for identifying people around the robot. People wear the tags, and the robot carries the tag reader. The active tag has a longer range of detection than the passive tag. But it needs a battery, and battery life is not so long. It depends on the use, but the battery needs to be frequently changed. On the other hand, the passive tag does not require a battery, but the detection range is about 10 mm. People need to put the tag on the tag reader. So, it is important to choose the type of tag according to the purpose and situation.

The most powerful sensors are laser scanners and cameras. Cameras are compact and easy to use, especially in an indoor environment. However, they are not good for an outdoor environment since the dynamic range for brightness is limited. If there are very bright areas and very dark areas in the same scene, they cannot recognize targets such as humans and robots. On the other hand, the laser scanner is quite stable to such change of brightness. The best way is to combine both sensors. If we use several laser scanners, we can track humans and robots stably. Then, we can use cameras for recognizing the face and facial expression based on the detected positions by the laser scanners.

The last sensor is the skin sensor used for the robot body. The robot interacts with people and needs to have soft and sensitive skin and skin sensors. Unfortunately, the development of the skin sensor is not so easy. Although it is quite an important sensor not only for robots but also other equipment and many researchers are tackling its development, we have not developed an ideal one. Sustainable development is needed.

These sensors form sensor networks in various ways, and the sensor network for the robots works in three ways:

1. Enhancement of the sensors on the robots
2. Monitoring events in the environment for safety
3. Realization of autonomous robots

For this role, we have to densely distribute sensors and precisely monitor both robots and humans. The sensors on the robots are not powerful and not intelligent since the robot cannot carry a sufficient number of sensors and enough computing resources. For this purpose, active and passive tags are used, for example.

For the second role of sensor networks, a powerful sensor network that consists of laser scanners and cameras is used. The sensor network can precisely track both robots and humans in real time. Therefore, it can detect dangerous situations involving robots and humans.

The third role of the sensor network is to make the robot autonomous. For this role, we have to use all sensors and densely and precisely monitor and record both robots and humans. The rich sensory information obtained by sensor networks enables us to develop autonomous robots that work in the real environment.

This section introduces sensor networks consisting of various sensors and discusses how they support both robots and humans.

5.2 Laser-Based Tracking of Human Position and Orientation Using Parametric Shape Modeling

Dylan F. Glas, Takahiro Miyashita, Hiroshi Ishiguro, and Norihiro Hagita

ATR Intelligent Robotics and Communication Laboratories,
2-2-2 Hikaridai, Keihanna Science City, 619-0288, Kyoto, Japan

ABSTRACT

Robots designed to interact socially with people require reliable estimates of human position and motion. Additional pose data such as body orientation may enable a robot to interact more effectively by providing a basis for inferring contextual social information such as people's intentions and relationships. To this end, we have developed a system for simultaneously tracking the position and body orientation of many people, using a network of laser range finders mounted at torso height. An individual particle filter is used to track the position and velocity of each human, and a parametric shape model representing the person's cross-sectional contour is fit to the observed data at each step. We demonstrate the system's tracking accuracy quantitatively in laboratory trials, and we present results from a field experiment observing subjects walking through the lobby of a building. The results show that our method can closely track torso and arm movements, even with noisy and incomplete sensor data, and we present examples of social information observable from this orientation and positioning information that may be useful for social robots.

Keywords: People tracking, particle filtering, motion analysis, human–robot interaction, laser-based tracking

5.2.1* Introduction

A new class of service robots is emerging, one in which social interaction is a fundamental aspect of a robot's performance. Experimental field trials have demonstrated the possibility of robots acting as museum guides [1], receptionists [2], classroom assistants [3], guides in shopping centers, and other social roles in everyday life. As the natural-language and gestural communication capabilities of these robots improve, people's expectations of the robots' interaction skills will commensurately increase and these robots will need to be responsive not only to speech, but to subtle cues of nonverbal communication as well.

Movement and positioning, for example, contain implicit information about a person's intentions, social relationships, mood, and status. A person's walking speed, trajectory, proximity to other people, and facing direction all provide information that can contribute to an understanding of social context.

Such knowledge could be used by service and communication robots to identify people who have lost their way or are in need of help, to stay out of the way of people in a hurry, to identify group leaders for guidance or sales applications, to understand when the robot is the center of attention

---

* This chapter is adopted from a previously published paper Dylan F. Glas, Takahiro Miyashita, Hiroshi Ishiguro, and Norihiro Hagita, *Laser-Based Tracking of Human Position and Orientation Using Parametric Shape Modeling*, in Advanced Robotics, Vol. 23, No. 4, pp. 405–428, 2009.

and when it is being ignored, to identify booths in an exhibition or exhibits in a museum that a person has missed, and for many other purposes.

Although a robot's on-board sensors can be used for some of these tasks, ubiquitous sensor networks can monitor larger areas and are subject to fewer size, power, and bandwidth restrictions. In many of our experiments and field trials, laser range finders are used for tracking people's positions as they are easier to install and less obtrusive than floor sensors, require far less processing than video tracking systems, and have a much higher precision and faster response time than RFID tracking or GPS.

To use these resources effectively, one goal of our research is to extract as much information as possible from this laser scan data. If nuances of a person's movement, such as the direction in which they are facing, can be extracted from the same laser scan data already used to determine their location, then information that is potentially useful for understanding social context will have been gained at no additional hardware cost.

In this paper, we present an algorithm we have developed for tracking people using a parametric shape model that includes arm positions and facing direction in addition to basic position tracking. The algorithms used in this system are described and quantitative results of a laboratory experiment to characterize the system's tracking accuracy are presented. A second experiment was conducted in the entrance lobby of an office building, to demonstrate the system's performance with multiple subjects in natural walking situations. Qualitative results from that experiment are presented, illustrating the system's effectiveness in tracking many people simultaneously and suggesting types of social information that can be observed in the tracking results. Finally, considerations concerning performance tuning and real-time operation of the system are discussed.

5.2.2 Related Work

Human tracking itself is not a new field, and many aspects of the problem have already been explored extensively. Like many of its predecessors, our system tracks people by using particle filters to estimate their position and velocity. Particle filters are a well-known tool in the robotics community, and have often been used in conjunction with laser scan data for the purposes of robot localization and mapping [4,5] as well as human tracking. A general overview of applications of particle filters in robotics can be found in [6].

Much of the human-tracking research to date has been based on leg tracking, for both mobile robotics [7,8] and environmental monitoring [9–11]. This has historically been motivated in part by the fact that many robots use laser sensors for obstacle avoidance and for that reason already have laser sensors mounted near the ground. However, their visibility is often limited by those same obstacles, making floor-level sensors a good choice for on-board robot systems, but less so for wide-area environment monitoring in cluttered spaces.

In our work, the laser sensors constitute an essential part of a ubiquitous sensor network used exclusively for human tracking in real environments. For this reason, it is important for the sensors to be mounted higher, above furniture and ground clutter. Thus, the sensors in our system are mounted at a height of 85–90 cm, where the arms and torso can be clearly observed.

Although less common than leg tracking, torso-level tracking is not without precedent in research. For example, Fod et al. created a system using a Kalman filter to track people's trajectories with waist-height laser scanners [12] and Almeida et al. developed a real-time torso-level laser-based human tracking system utilizing particle filters [13]. These systems, however, were focused specifically on position tracking, whereas our work is concerned with observing body orientation and pose in addition to position.

5.2.3 Position Tracking

Our algorithm was developed to track both human position and orientation. The strategy of this algorithm is to first estimate each person's position using a particle filter and then to fit a shape model, representing the person's body orientation and arm positions, to the observed contour data.

Our initial approach to this problem had been to calculate both position and orientation using the particle filter. This resulted in an unacceptably slow system for our real-time applications. However, we observed that a majority of the computation time for each particle was being spent on orientation calculations.

In fact, the edge-based calculations used for orientation are not particularly well suited for use in a particle filter. For position, calculations are efficient because their likelihood distributions are stable over time (regions are clearly defined and change slowly), relatively smooth in space, and easy to calculate from raw sensor data. Edge-based likelihood distributions are more complex to calculate, not stable over time (the number and placement of detected points can change rapidly between frames with a great deal of randomness), nor are they smooth in space, as the best-fit orientation can change wildly over even small variations of the assumed center position. It is thus difficult to obtain a meaningful average orientation value over a scattered set of particles.

In our technique, the orientation calculations are highly dependent upon position, but the position calculations do not depend on orientation. Thus, the orientation calculations can be removed from the particle filter and performed after the position estimate is evaluated. Having done so, at each time step we need only calculate orientation once for each particle filter, rather than once for each particle. In addition, by removing variables from the particle filter we are able to reduce its dimensionality, consequently reducing the number of particles necessary for accurate tracking. By separating the calculations into a two-step process, we are thus able to dramatically increase real-time performance. More details on this topic can be found in Section 5.2.7.2.

5.2.3.1 Detection and Association

A common problem in tracking is the association between detected features and objects being tracked. In our algorithm, each person is tracked by a single particle filter. Doing so enables these feature–object associations to be handled implicitly by the particle filters, which follow the detected features over time. Thus, explicit feature–object associations only need to be made when creating new particle filters for previously untracked humans.

To identify new humans, the raw data are segmented at every time step, to extract continuous segments of foreground data roughly corresponding to expected human widths. Clusters of these patterns are grouped together and flagged as human candidates. Candidates coinciding with humans already being tracked are removed from the list and those remaining are propagated to the next time step, where they are merged with the candidates detected during that step. If a human candidate survives beyond a threshold number of time steps, it is considered to be a valid detection, and a new particle filter is assigned to that location, initialized with the position and velocity of the human candidate it replaces.

The removal process is much simpler than the addition process. When the particles within a filter spread out beyond a defined dispersion threshold or when their average likelihood value goes below a defined probability threshold, that particle filter is assumed to no longer be tracking a human and it is removed.

5.2.3.2 Particle Filtering

A key component of our tracking algorithm is the particle filter, the basic principles of which will be very briefly explained here. For a more in-depth explanation, [14] provides a thorough treatment of particle filters and many other state estimation techniques.

Particle filtering is a method of estimating the state x_t of a system by using a cloud of "particles," each of which represents a hypothesis about that state. The following four-step procedure is performed at each iteration of a particle filter:

i. *Update.* The state of each particle is updated by applying an internal motion model, reflecting the dynamics of the system, to the previous state estimate. The motion model used in our work is described in Section 5.2.3.4.

ii. *Assign weights.* Particles are assigned weights representing their relative likelihoods according to a likelihood model. The likelihood model provides an approximation of the conditional probability

$p(z_t | \mathbf{x}_t^{[m]})$ for particle m ($m = 1, \ldots, M$), and measurement vector z_t taken at time step t of the particle filter. Our likelihood model is described in Section 5.2.3.5.

iii. Estimate state. An estimate of the state is then calculated, generally as a weighted average of the states of the particles.

iv. Resample. Particles are removed or propagated based on their weights to produce a new set of particles which more accurately reflects the true state of the system. Several resampling techniques exist; our system uses the sampling importance resampling technique [15].

In this way, the cloud of particles converges on the most likely state and follows it over time.

5.2.3.3 State Model

The state vector tracked by the particle filter consists of four variables: x, y, v, and θ. The variables x and y represent the position of the human being tracked. Although the speed v and direction θ of motion could be calculated a posteriori from the position data, these variables are included in the state and updated at every step to enable the person's position to be projected forward through time for more accurate tracking. These variables are used in the motion model described below.

5.2.3.4 Motion Model

At every update of the particle filter, each particle is propagated according to a motion model. The purpose of this motion model is to approximate the probability of a state $\mathbf{x}_t$ based on the previous state $\mathbf{x}_{t-1}$.

As has been observed in [16], the modeling of human motion presents difficulty because it is neither Brownian in nature nor can it be modeled as a smooth linear function, since people may stop or change direction abruptly. Thus, as a compromise between the two, a Gaussian noise component is added to each particle's v and θ values to capture the randomness of human motion. We then propagate the (x, y) motion linearly according to the resultant v and θ values of the particle.

5.2.3.5 Likelihood Model

The purpose of the likelihood model is to approximate the value of $p(z_t | \mathbf{x}_t^{[m]})$. In this case, the measurement vector z is an array of raw sensor range measurements. An effective likelihood model must provide a robust likelihood estimate in spite of noisy sensor data, partial and full occlusions, and the irregular and varying shapes of human bodies.

Laser scan data provide two qualitatively distinct types of information useful for estimating human positions: occupancy information, indicating whether a certain point is occupied or empty; and edge information, indicating a contour that may correspond with the edge of a detected object. Figure 5.1 illustrates the distinction between these two kinds of information.

To determine likelihood values from the raw sensor data, it is first necessary to create a background model. Our system uses an adaptive background model that is updated over time to determine the best estimate of the true background distance. Occupancy likelihood is then determined by dividing the world into three regions: "open," "shadow," and "unobservable." The unobservable region is beyond the background model for that sensor and thus can contribute no information. The open region has been observed by the sensor to be unoccupied, and the remaining space is considered shadow. Note also that every shadow region lies behind an "edge."

The likelihood model used to compute $p(z_t|\mathbf{x}_t^{[m]})$ is expressed in (5.1) and (5.2), and includes components reflecting both occupancy and edge information:

$$p\left(z_t\middle|\mathbf{x}_t^{[m]}\right) = \frac{1}{n_{sensors}} \sum_{i=1}^{n_{sensors}} p_i\left(z_t\middle|\mathbf{x}_t^{[m]}\right) - p_{collocation} \tag{5.1}$$

$$p_i\left(z_t\middle|\mathbf{x}_t^{[m]}\right) = \begin{cases} p_{shadow} + p_{edge}\left(z_t\middle|\mathbf{x}_t^{[m]}\right) & |shadow \\ p_{open} & |open \end{cases} \tag{5.2}$$

For a point in a shadow region (strictly speaking, we consider only those regions wide enough to contain a human), the likelihood in (5.2) is calculated as the sum of a constant value p_{shadow} and a likelihood $p_{edge}(z_t|\mathbf{x}_t^{[m]})$, calculated as a normal distribution centered upon a point located one approximate human radius behind the observed edge (in our calculations, a value of 25 cm was used). This reflects the fact that people are highly likely to be found just behind an observed edge, yet can plausibly exist anywhere in a shadow region (e.g., the occluded person in Figure 5.1).

For a point in an open region (or in a shadow region too narrow to contain a human), the likelihood is theoretically zero, but for reasons described below is set to a small but nonzero constant value p_{open}. In this case, edge information is irrelevant.

Finally, in (5.1), these likelihood values are averaged across all $n_{sensors}$ sensors for which the proposed point lies within the sensor's open or shadow range, that is, not "unobservable" to that sensor. To prevent two particle filters from tracking the same human, a value $p_{collocation}$ is subtracted from this result. Its value is calculated as a sum of normal distributions surrounding each of the other humans, based on the list of human positions from the previous time step.

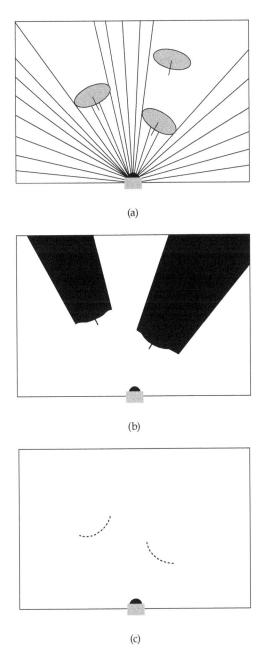

(a)

(b)

(c)

FIGURE 5.1
A typical single-sensor laser scan. (a) The positions of humans relative to the scanner can be seen. (b) Occupancy information. (c) Edge information.

5.2.3.5.1 Error Tolerance

In an ideal system, the open regions could be assigned a likelihood value of zero. However, in real systems there are many possible sources of error, such as calibration errors (the exact position and angle of each sensor may not be properly calibrated, leading to imperfect alignment of shadow regions), measurement errors (some textures of clothing cause noisy sensor readings and thus apparent gaps in people's bodies), timing synchronization errors (sensor data feeds are sent in real time over a network and may arrive asynchronously, causing old data to be mixed with new) and hardware or transmission errors (which produce occasional bursts of sensor noise). The binary discretization of space into open and shadow regions is thus a slightly imperfect representation of reality. Consequently, we set the likelihood of open regions to a small but nonzero value p_{open}. This adds a small amount of resilience to the system, allowing particles to survive outside of the shadow regions for a short time in order to provide smoother performance with respect to such sources of error. This does not destabilize the particle filter since the likelihoods of these particles are substantially lower and particles lying outside of the shadow regions for too long will naturally be culled in the resampling process.

5.2.4 Orientation Estimation

Our algorithm for calculating a person's orientation uses a parametric shape model, which we describe in Section 5.2.4.1. An angular array representation, presented in Section 5.2.4.2, is used to store laser scanner data as a set of edge distances. As a tool for our calculations, an empirical distribution of expected distances for such an array, relative to the person's forward-facing direction, was generated based on laboratory motion-capture data. We describe the derivation of this distribution in Section 5.2.4.3.

The computation itself consists of first determining a rough estimate of body orientation, described in Section 5.2.4.4, based on the observed contour shape and the empirical distance distribution mentioned above. The second step, explained in Section 5.2.4.5, is to determine the individual arm angles, based on this rough estimate. The arm angles are then used to generate a refined estimate of orientation. Finally, Section 5.2.4.6 presents a technique for reducing accidental 180° reversals by considering motion direction and velocity.

5.2.4.1 Theoretical Shape Model

Large variations in cross-sectional contour shape were observed between subjects. This is due in part to individual differences in body shape and also to differences in height. For example, arm motion is more pronounced for taller subjects, and their arms sometimes disappear if their hands briefly swing out of the scan plane.

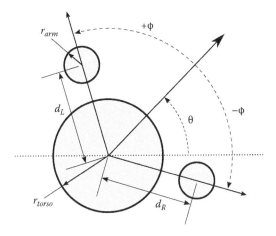

FIGURE 5.2
Our three-circle model, with the six variable parameters indicated.

Clothing also affects contour shape. For example, a loose shirt or a heavy coat can make a person's torso appear unusually large or asymmetrical, as can a backpack or purse.

Taking these factors into consideration, the amount of variation between subjects makes it difficult to develop a precise, yet generalizable, model. Thus, a simple three-circle model was used for determining body orientation.

Our model is illustrated in Figure 5.2. A central, large circle represents the person's torso, and two smaller circles represent the arms. This model has six parameters which can be varied to best match a subject's cross-sectional body contour.

The parameters describing the state of this model are summarized in Table 5.1. The two parameters of primary interest to us are θ and ϕ. The other parameters are held constant for this application, although they can be estimated from the data if necessary.

TABLE 5.1

Model Parameters

Parameter	Description
θ	Average direction of body orientation
ϕ	Arm separation angel:
	$\phi_L = \theta + \phi$ for left arm
	$\phi_R = \theta - \phi$ for right arm
d_L	Distance of left arm from body
d_R	Distance of right arm from body
r_{arm}	Arm radius
r_{torso}	Torso radius

We have designated θ to represent the angle midway between the two arms. When a subject is standing still, this coincides with the direction of torso orientation. While the subject is walking, the swinging of the arms and torso causes θ to oscillate around the direction of motion.

The parameter ϕ represents the angle of separation between each arm and the center angle designated by θ. This tends not to vary far from 90°, as the arms swing in alternate directions during walking.

5.2.4.2 Radial Data Representation

For these calculations, we need a way to represent two-dimensional edge data in a consistent way for analysis. To achieve this, the information contained in these points is mapped to an angular array of distances. Distance values from the body center to the detected edge points are stored in an array of bins that represent an angular discretization of the space surrounding the estimated human position. For each angular division, the distance to the furthest observed data point within 50 cm of the estimated human position is stored in that bin. Figure 5.3 illustrates such array representations of both the ideal shape model and a set of actual shape data (a linear representation of such an array is shown below in Figure 5.5a).

5.2.4.3 Empirical Distance Distribution Model

A predictive distribution of radial distances is also needed for these calculations. An empirically derived predictive distribution function representing average expected distance values as a function of angular deviation from θ was constructed from the laser scan and motion capture data gathered in

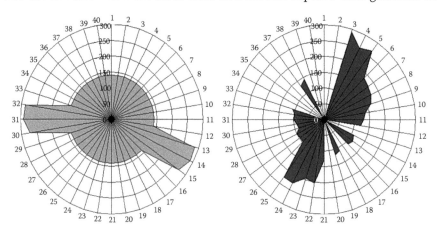

FIGURE 5.3
Examples of populated radial arrays. (Left) Radial array reflecting an ideal human shape model. (Right) Radial array populated with observed sensor data.

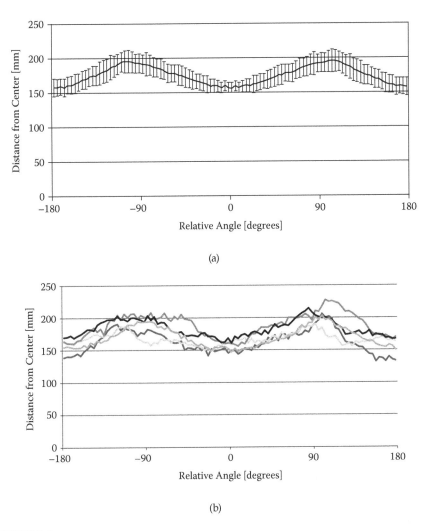

(a)

(b)

FIGURE 5.4
(a) Predictive arm distribution filter showing standard deviation error and (b) raw data used to derive the filter.

the laboratory trials described in Section 5.2.5. This distribution function is shown in Figure 5.4a.

Two minutes of laser scan and motion capture data were recorded for each of five subjects. Each subject's angle at each time step was computed using the motion capture system, and a radial accumulator with 100 divisions (3.6° each) was populated with the laser scan data for that time step, oriented relative to that angle. This distance data was collected over approximately 4500 time steps and averaged to determine an expected distance distribution function for each subject. These distribution functions are shown in Figure 5.4b.

Next, the data distributions were averaged between subjects. The resultant function was still somewhat noisy and asymmetrical. Making the assumption that this distribution should be symmetrical (and if there is a physiological reason for the asymmetry, to eliminate any bias based on handedness), the mirror images of the subjects' data distributions were also included in the average. Figure 5.4a shows the standard deviation error bars for this combined distribution. The resultant distribution was then smoothed using a sliding three-point window to reduce remaining noise. Finally, a constant offset was subtracted from the filter and normalized, steps which do not alter its effectiveness as a convolution filter.

5.2.4.4 First-Pass θ Determination

The strategy for the first approximation of θ involves two radial arrays. The first is populated with the actual observed distance of data points from the body center, with the angular divisions corresponding to absolute angles. The second array holds the expected distribution of distances derived in Section 5.2.4.3, where the angular divisions represent angles relative to θ, the person's forward direction. By convolving these arrays with each other, we can compute a goodness-of-fit function between the predicted distribution and the observed distribution, as a function of θ. The maximum point of that function is the point where the observed data best fits with the expectation model and is thus a good first-pass estimate for θ.

To begin, we need to construct an approximate model of the actual shape profile, beginning with the radial array shown in Figure 5.5a. There will nearly always be angular divisions in the radial array with no points in them. Since we have no knowledge of the actual distances of these points, we set those bins to the average value across all occupied bins, to produce a model with no gaps, as shown in Figure 5.5b. This same array will be normalized and used later as a probability distribution function for arm positions, as explained below.

This distribution, shown in Figure 5.5c, is convolved with the data array shown in Figure 5.5b to generate a function representing the goodness of fit between the observed data and the predicted data distribution. The maximum point of the resultant distribution indicates the θ value that gives the best match between the empirical distribution and the observed data.

One challenge in this determination of θ lies in the near symmetry of the human body. Although the expected value of the arm angles is less than 180°, the observed distribution and its 180° mirror image overlap significantly. Thus, particularly with noisy and incomplete data, it is possible that the best-fit angle is actually rotated 180° from the true θ direction. To stabilize this variable, the secondary maximum in the θ likelihood function is designated as a second θ candidate. The angular distance from the previous θ estimate to the two new θ candidates is compared, and the nearest neighbor selected as the first-order θ approximation. Correction of these reversals is discussed in Section 5.2.4.6.

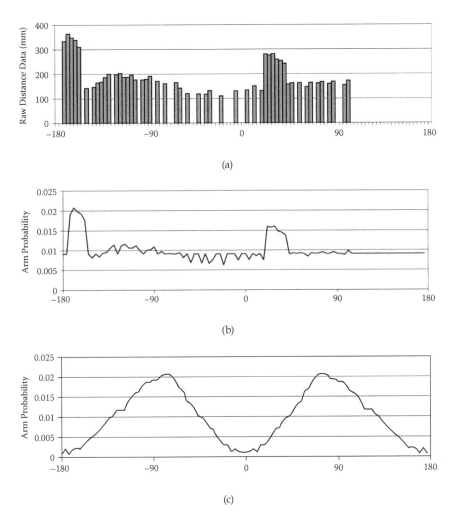

FIGURE 5.5
Intermediate steps in arm angle determination. (a) Maximum observed distance values in raw
data array. (b) Interpolated shape profile/arm angle probability distribution. (c) Empirically
derived theta-centric distance distribution. (d) Result of convolution with distance distribution.
(e) Masking functions for left and right arms. (f) Probability functions for left and right arms.

5.2.4.5 Second-Pass θ Determination

Using this rough θ estimate, the next step is to determine the arm angles ϕ_L
and ϕ_R, which will be used for determining the final θ estimate. For this step,
it is necessary to derive a probability distribution function (PDF) for the arm
positions from the observed data.

For this purpose, the shape profile model derived in the previous step
can be used as a rough approximation of the arm position PDF, as it exhibits many of the essential features of such a distribution. For example, the

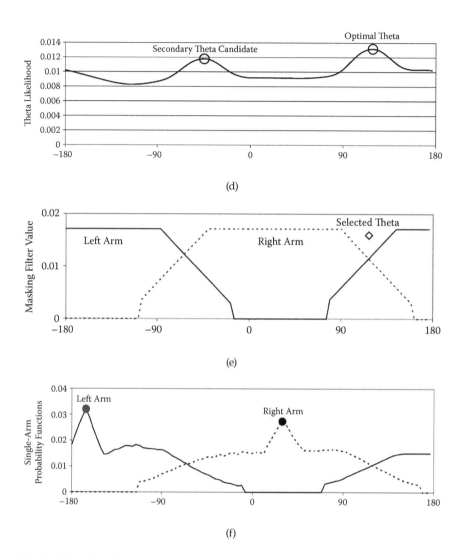

(d)

(e)

(f)

FIGURE 5.5 (continued)

presence of distant points indicates a high likelihood that an arm is in that direction. Likewise, the presence of closer-than-average points indicates a low likelihood of an arm being in that direction. Several points observed in a row give a higher confidence estimate than a single point, high or low, and points with no data provide no information about the presence or absence of an arm. All of these features are found in both a theoretical PDF for arm distribution as well as the data array derived above. Thus, by normalizing that array, we obtain a rough approximation of that PDF.

The arm probability distribution in the radial array is then masked into two 180° regions by using trapezoidal masking filters on either side of the selected θ direction as shown in Figure 5.5e (trapezoidal rather than rectangular masks were used for stability). These masks are multiplied with the data array from Figure 5.5b to generate the two probability distributions shown in Figure 5.5f. The peaks of these distributions are used as estimations of the left and right arm angles ϕ_L and ϕ_R, respectively. A refined estimate of θ is then calculated as the midpoint between these angles.

Note that at this point, if desired, the shape profile can be revisited to calculate parameters such as d_L, d_R, r_{arm}, and r_{torso}. However, this step is not necessary if θ is the only parameter of interest.

5.2.4.6 Correction of Reversals

One of the greatest difficulties in determining the person's facing direction lies in resolving the 180° ambiguity between forward and backward orientation. The human shape is nearly symmetrical, and even by eye it is sometimes quite difficult to discern front and back from laser scan data alone.

To resolve this ambiguity, we utilize the assumption that motion direction generally tends to coincide with the forward orientation direction. We verified this assumption quantitatively using the data recorded in the trials described in Section 5.2.5.

By running the basic human-tracking program without any reversal correction, we generated a dataset of human positions and orientations. Reversals (defined as periods in which the directional error was greater than 90°) were identified by comparing these results with the ground-truth data from the motion capture system. A velocity distribution was then computed for each set of data points. The results of this analysis are illustrated in Figure 5.6.

An examination of this velocity distribution reveals that retrograde motion at low velocities is common, probably due to a combination of actual motion, noise, and tracking lag of the particle filter; however, higher retrograde velocities (above 500 mm/s) are almost nonexistent. Thus, any retrograde motion larger than a threshold speed of 500 mm/s is interpreted as a reversal and corrected. A time-averaged velocity estimate is used to minimize the influence of noise.

5.2.5 Laboratory Performance Analysis

We performed an experiment in our laboratory to verify the accuracy of the human tracking system, and to gather empirical data to refine the reversal detection and θ approximation functions in our tracking algorithm.

5.2.5.1 Setup and Procedure

We used a Vicon motion capture system to measure the accuracy of our laser tracking system. The Vicon system uses several infrared cameras to track

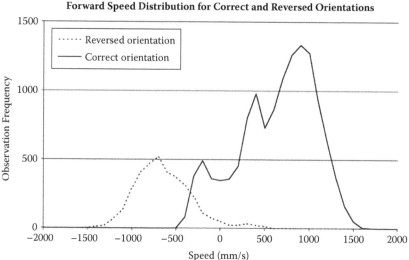

FIGURE 5.6

Distribution of forward velocity. The forward component of the velocity vector was calculated for each time step, and a frequency histogram was computed using bin sizes of 100 mm/s. Nearly every observation with a forward velocity component below −500 mm/s was the result of a reversed direction estimate.

reflective markers with an accuracy of 1 mm at a frequency of 60 Hz. Four SICK LMS-200 laser scanners were used, each set to a maximum range of 8 m, a distance resolution of 10 mm, an angular range of 180°, an angular resolution of 0.5°, and a scan frequency of 37.5 Hz.

The space used for our experiment was an area of 4 m × 4 m with the four laser scanners situated outside the center of each edge of the square. The scan plane for each laser scanner was located at a height of 85 cm from the ground. Additionally, numbered markers were placed on the floor, as depicted in Figure 5.7.

Five subjects were instructed to walk a series of patterns within the square. First, they stood in the center of the square and turned in a circle, stopping at each of the four cardinal directions for 2 s. Second, they walked figure-of-eight patterns, touching each of the numbered markers in order, twice. Third, they walked in a circular path inside the square, twice clockwise and twice counterclockwise. Finally, they walked randomly within the square until a total of 2 min had elapsed.

Each subject wore four reflective markers for the Vicon system. One marker was placed on the outside of each wrist, one on the subject's sternum, and one in the middle of the subject's back.

Raw data from each of the laser scanners were recorded, and the human tracking algorithm was executed offline.

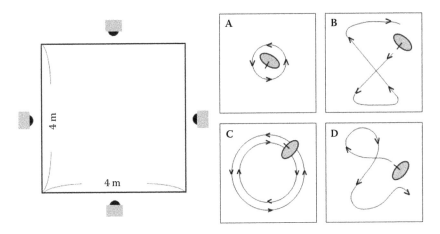

FIGURE 5.7
Floor layout for our laser tracking validation tests. Subjects were observed by four laser range finders while walking several patterns within a 4 m × 4 m square.

5.2.5.2 Results

To compare the motion capture data with the laser tracking data, the midpoint between each subject's sternum and back markers was used as an estimate of the subject's body center. The absolute positional error (in the x, y plane) and absolute angular error between the laser tracking data and the motion capture data were then calculated for every time step in the laser-based tracking data (Figure 5.8).

The average positional error over all five subjects was 4.6 ± 2.7 cm, and the average angular error was 8.2 ± 13.8°. During the 10 min of tracking, there were nine brief 180° reversal errors. One of these lasted for 2.2 s, and all others were automatically corrected within 0.2 s. The average error with those intervals excluded from the data was 7.4 ± 7.9°.

5.2.6 Natural Walking Experiment

Although the trials in our motion capture room provided useful data for verifying the system's accuracy, it is difficult to simulate natural human walking motion in such a restricted space.

To verify that the system could also work with natural walking data, we ran several trials in an open lobby, roughly 19 m long and 8 m wide. Experimental subjects were instructed to walk through the area several times under a number of different conditions, for example, individually, in groups, wandering aimlessly, walking purposefully, making U-turns, and stopping to ask for directions.

Raw data from a network of six laser range finders monitoring this area were recorded for each trial, which we processed offline to determine human positions.

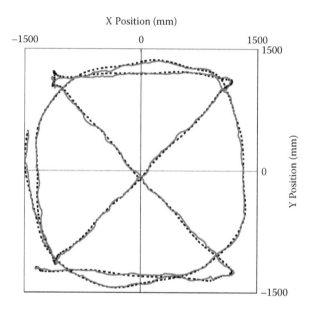

FIGURE 5.8

Tracking example from walking data. The dashed line represents ground-truth data from the motion capture system, and the solid line represents laser tracking data.

5.2.6.1 Setup and Procedure

The area of interest in our experimental environment was a space within the lobby roughly 19 m long and 8 m wide. We used six SICK LMS-200 laser scanners, set to scan an angular area of 180° at a resolution of 0.5°, covering a radial distance of 8 m with a nominal system error of ±20 mm, providing readings of 361 data points every 26 ms. These were placed around the periphery of the experimental area such that every point within the area of interest would be covered by at least two sensors, to minimize occlusions.

The sensors were mounted at a uniform height of 90 cm, slightly above waist level for most subjects. Tables, benches, and a small mobile robot were also placed within the walking area, but all of these were below 90 cm and thus not visible to the laser scanners.

Twelve adults participated as subjects in this experiment, although at any given time only a subset of the group was walking within the sensor area. Six trials were conducted, and a total of 172 min of raw sensor data was collected.

5.2.6.2 Results

Two aspects of the results of this experiment will be considered here. The first is the accuracy of our method in tracking the subjects' motions, and the second is the ability to interpret this data in terms of actual body language and behavior.

5.2.6.2.1 Tracking Individuals

Quantifying the accuracy of this tracking technique is challenging due to the lack of more precise measurement techniques to establish a ground truth for evaluation. A side-by-side visual comparison of the raw data with the model-based estimate is perhaps the most effective indicator of the tracking accuracy.

Figure 5.9 shows raw data from five frames taken during the course of a single stride and compares them with the model-based estimates for those time frames. Note that the swinging of the arm is clearly visible from the data and that the model follows this movement closely.

Another indicator of the tracking accuracy of our technique is the resolution of movement that is visible over time. Figure 5.10 shows a sample path walked by one of the subjects during our experiment. The variations in θ due to the swinging of the arms and torso with each stride are quite clearly visible, with little noise present. The more subtle change in angle as the subject walks around a curving path is also quite clearly visible from the data.

These tracking results were then visually compared with video recorded during the experiment. The subjects' arm-swinging motions were observed to match with the data. The subject's torso rotations were not as exaggerated as the variations of θ in our model, which suggests the possibility that modeling the motion of the arms during walking may offer a better estimate of torso orientation.

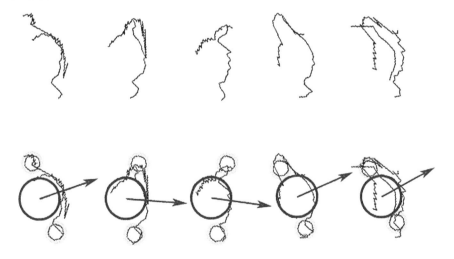

FIGURE 5.9
Example of arm and torso movement during a single stride. (Top) Five frames of raw data from laser scanners taken at 320 ms intervals. (Bottom) Corresponding human shape model positions for each frame.

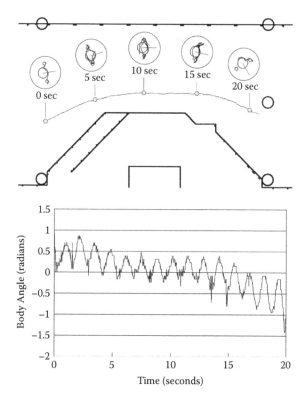

FIGURE 5.10
Body angle tracking during 20 s of walking motion. (Top) An overview of the walking path in our lobby experiment shows the subject's walking path, as well as close-up views of the subject's body position at several points along the path. (Bottom) Observed body angle variations (in room-centric coordinates). Periodic oscillations due to the natural arm-swinging motion during each stride are clearly visible.

Interestingly, our tracking results for one trial indicated an asymmetry of motion, with one arm moving much more than the other. Inspection of the video revealed that this was not a tracking error at all, but an idiosyncrasy of the subject's walking style—an observation that suggests the possibility of using the information in this model for identifying individuals or making inferences about personality or mood.

5.2.6.2.2 Observing Interactions

In addition to the model's tracking accuracy, it is important to consider what information can be observed regarding groups of people in social situations.

Figure 5.11 shows three scenes from our experiment. In the top scene, two subjects are seen walking together. The model correctly shows that they are walking side by side, facing each other slightly. It is possible that the relative directions in which people face while walking together might include information about their social relationship.

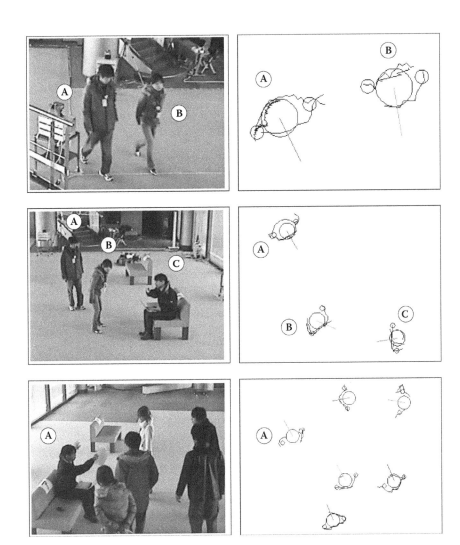

FIGURE 5.11
Scenes from the experiment.

In the center scene, one of the two subjects is asking a third subject for directions. The model clearly shows the social situation, in which Subjects A and B are focusing their attention on Subject C. Subject A is standing back at a respectful distance, which seems to imply that A and B are not part of the same group or perhaps that their relationship is very formal.

The bottom scene illustrates the tracking of a group of subjects. Again, the group dynamic is apparent, in that all of the subjects are listening to instructions from Subject A. (Note that the model is unable to correctly determine the direction of Subject A because he is sitting and holding his arms in an unusual position.)

All three of these examples illustrate information that could not have been determined from location alone, and they suggest many possible types of social information that may be observable from this data.

5.2.7 Discussion

5.2.7.1 Performance Tuning

Many variables affect the performance of the system in terms of operating speed, position and angle accuracy, smoothness of motion, and false or missed detections. By reducing the velocity noise added during the motion model updates, for example, higher positional accuracy and smoother trajectories can be attained, but the particle filter becomes less able to follow trajectories that change abruptly.

The number of particles is another variable. If a large number of particles is used, the particle cloud's trajectory stabilizes and becomes smoother, but this comes at the cost of increased reaction delay and increased computation time. Our algorithm uses the technique of KLD sampling [17] to adapt the number of particles based on the density of their distribution down to a fixed minimum limit.

5.2.7.2 Real-Time Operation

Although the results presented in this chapter were generated offline, this tracking software has primarily been developed for use with real-time data streams. Using this software in a real-time system raises the critical issue of processing speed. If the time required to process the data for one time step exceeds the sampling interval of the sensors, then data will be lost, and tracking accuracy will begin to decrease.

Here we present a performance analysis using a Windows XP system with a 2.4 GHz Intel Pentium 4 processor and 1 GB of RAM. The tracking software was implemented in Java and executing using a Java 6 Virtual Machine. The tracking analysis was performed on a 4.5 min data sample from a shopping center, during which between 3 and 18 people were tracked simultaneously. Six sensors were used in this experiment, with a frequency of 37.5 Hz, that is, a sampling interval of 26.6 ms. A minimum of 50 particles were used for each person.

To illustrate the importance of the speed improvement gained by performing the orientation calculations separately from the particle filter, Figure 5.12 compares our system's performance against an algorithm in which orientation calculations are integrated with the position calculations within the particle filter.

This performance comparison illustrates two key points. The first point is that even with the relatively slow Pentium 4 machine used here, it can be expected that 10–12 people can be tracked without any loss of data; that is,

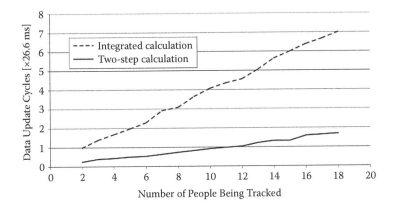

FIGURE 5.12
Variation of average computation time with number of humans being tracked. These results indicate that 10–12 people can be tracked before computation time exceeds the sensor sampling interval of 26.6 ms.

the tracking calculations can be completed within one data update cycle. With 18 people, incoming data would be dropped, but every second data frame would still be processed.

The second point is that, as stated in Section 5.2.3, the orientation computations are not very well suited for integration with the particle filter. Figure 5.12 shows that the integrated algorithm requires about four times the computation time of the two-step algorithm. In other words, the improved efficiency of the two-step algorithm enables four times as many people to be tracked at once. To address the question of whether the choice of algorithm affects tracking accuracy, we repeated the analysis from Section 5.2 using the integrated algorithm. Results were substantially worse than with the two-step algorithm. First, many more reversals were observed with the integrated algorithm. Even correcting for the reversals, the average angular error was still 25.2°, as opposed to 7.4° for the two-step algorithm. This was most likely due to the issues stated in Section 5.2.3, such as the nonsmooth likelihood model and high sensitivity to position error.

5.2.7.3 Future Work

The next step in this research is to use the generated position and orientation data to improve robotic applications. Techniques should be developed for analyzing a person's trajectory through a given environment to learn about that person's intentions. Information about the directions in which people in a group are facing and their relative standing or walking positions may be helpful in identifying social rank within that group. Trajectory and orientation data might be useful in identifying people in a crowd who are interested in talking with the robot or who have lost their way and need guidance.

Another possible area for future research is the addition of anatomically based physical dynamics. Rather than simply modeling motion using a geometric circular model, incorporation of arm swinging and stride motion into the model could provide much more stable and accurate results. Currently, the system is able to extract a person's torso direction, which has been observed to oscillate from left to right while walking. A more detailed dynamic model could incorporate walking speed and rhythm to determine an even better estimate of the person's direction of attention.

Finally, the integration of this system with other tracking technologies, such as a leg-based laser tracking system, could provide a very robust estimate of a person's pose and enable the interpretation of more subtle expressions of body language.

5.2.8 Conclusions

We have developed a system in which a network of laser range finders is used for tracking the positions and orientations of people.

Comparison with results from a motion capture system verified the position accuracy to be 4.6 ± 2.7 cm and the orientation accuracy of to be 7.4 ± 7.9° (excluding 180° reversals). The system is expected to perform without performance degradation while tracking 10–12 people in real time on a Pentium 4 Windows PC.

This human tracking system has already been used extensively for providing ground-truth data and tracking humans in several experiments and field trials. The system is also actively being used as a platform for extracting useful social information from human movement for social robotics applications.

References

1. W. Burgard, A. B. Cremers, D. Fox, D. Hähnel, G. Lakemeyer, D. Schulz, W. Steiner, and S. Thrun, The interactive museum tour-guide robot, in: *Proc. 15th Natl. Conference on Artificial Intelligence*, Madison, WI, pp. 11–18 (1998).
2. R. Gockley, A. Bruce, J. Forlizzi, M. Michalowski, A. Mundell, S. Rosenthal, B. Sellner, R. Simmons, K. Snipes, A. C. Schultz, and J. Wang, Designing robots for long-term social interaction, in: *Proc. IEEE/RSJ Int. Conf. on Intelligent Robots and Systems*, Edmonton, pp. 2199–2204 (2005).
3. T. Kanda, T. Hirano, D. Eaton, and H. Ishiguro, Interactive robots as social partners and peer tutors for children: a field trial, *Human Compub. Interact.* **19**, 61–84 (2004).
4. F. Dellaert, D. Fox, W. Burgard, and S. Thrun, Monte Carlo localization for mobile robots, in: *Proc. IEEE Int. Conf. on Robotics and Automation*, Detroit, MI, pp. 1322–1328 (1999).

5. M. Montemerlo, S. Thrun, and W. Whittaker, Conditional particle filters for simultaneous mobile robot localization and people-tracking, in: *Proc. IEEE Int. Conf. on Robotics and Automation*, Washington, DC, pp. 695–701 (2002).

6. S. Thrun, Particle filters in robotics, in: *Proc. Uncertainty in Artificial Intelligence*, Edmonton, AB, pp. 511–518 (2002).

7. D. Schulz, W. Burgard, D. Fox, and A. B. Cremens, People tracking with mobile robots using sample-based joint probablistic data association filters, *Int. J. Robotics Res.* **22**, 99–116 (2003).

8. A. M. Villagrasa, People tracking for a personal robot, Master's thesis, Royal Institute of Technology, Stockholm (2005).

9. A. Brooks and S. Williams, Tracking people with networks of heterogeneous sensors, in: *Proc. Australasian Conf. on Robotics and Automation*, Brisbane, pp. 1–7 (2003).

10. J. Cui, H. Zhao, and R. Shibasaki, Fusion of detection and matching based approaches for laser based multiple people tracking, in: *Proc. IEEE Computer Society Conf. on Computer Vision and Pattern Recognition*, New York, pp. 642–649 (2006).

11. H. Zhao and R. Shibasaki, A novel system for tracking pedestrians using multiple single-row laser range scanners, *IEEE Trans. Syst. Man Cybernet.* **35**, 283–291 (2005).

12. A. Fod, A. Howard, and M. J. Mataric, Laser-based people tracking, in: *Proc. IEEE Int. Conf. on Robotics and Automation*, Washington, DC, pp. 3024–3029 (2002).

13. A. Almeida, J. Almeida, and R. Araujo, Real-time tracking of moving objects using particle filters, in: *Proc. IEEE Int. Symp. on Industrial Electronics*, Dubrovnik, pp. 1327–1332 (2005).

14. S. Thrun, W. Burgard, and D. Fox, *Probabilistic Robotics*. MIT Press, Cambridge, MA (2005).

15. N. J. Gordon, D. J. Salmond, and A. F. M. Smith, Novel approach to nonlinear/non-Gaussian Bayesian state information, *Radar Signal Process. IEE Proc. F* **140**, 107–113 (1993).

16. A. Bruce and G. Gordon, Better motion prediction for people-tracking, in: *Proc. IEEE Int. Conf. on Robotics and Automation*, New Orleans, LA (2004).

17. D. Fox, KLD-sampling: Adaptive particle filters, *Adv. Neural Inform. Process. Syst.* **14**, 713–720 (2001).

5.3 Super-Flexible Skin Sensors Embedded on the Whole Body, Self-Organizing Based on Haptic Interactions

Tomoyuki Noda, Takahiro Miyashita, Hiroshi Ishiguro, and Norihiro Hagita

ABSTRACT

As robots become more ubiquitous in our daily lives, humans and robots are working in ever-closer physical proximity to each other. These close physical distances change the nature of human–robot interaction considerably. First, it becomes more important to consider safety, in case robots accidentally touch (or hit) the humans. Second, touch (or *haptic*) feedback from humans can be a useful additional channel for communication, and is a particularly natural one for humans to utilize. Covering the whole robot body with malleable tactile sensors can help to address the safety issues while providing a new communication interface. First, soft, compliant surfaces are less dangerous in the event of accidental human contact. Second, flexible sensors are capable of distinguishing many different types of touch (e.g., hard vs. gentle stroking). Since soft skin on a robot tends to invite humans to engage in even more touch interactions, it is doubly important that the robot can process haptic feedback from humans. In this chapter, we discuss attempts to solve some of the difficult new technical and information processing challenges presented by flexible touch sensitive skin. Our approach is based on a method for sensors to self-organize into sensor banks for classification of touch interactions. This is useful for distributed processing and helps to reduce the maintenance problems of manually configuring large numbers of sensors. We found that using sparse sensor banks containing as little as 15% of the full sensor set, it is possible to classify interaction scenarios with accuracy up to 80% in a 15-way forced choice task. Visualization of the learned subspaces shows that, for many categories of touch interactions, the learned sensor banks are composed mainly of physically local sensor groups. These results are promising and suggest that our proposed method can be effectively used for automatic analysis of touch behaviors in more complex tasks.

5.3.1* Introduction

Robots are becoming more ubiquitous in our daily lives [1–4], and humans and robots are working in ever-closer physical proximity to each other. Due to this proximity, there is increased potential for robots to inadvertently harm users. Physical nearness also increases the need for robots to be able to interpret the meaning of touch (or *haptic*) feedback from humans. Covering the whole robot body with malleable tactile sensors allows us to address both of these concerns. First, soft, compliant surfaces are less dangerous in the event of accidental human contact [5]. Second, flexible sensors are capable of distinguishing many different types of touch (e.g., hard vs. gentle stroking). This is important, as soft skin actually invites more natural types of touch interaction from humans, so it is critical that the soft surfaces of robots be touch sensitive.

To extract information about humans' physical contact with robots, the distribution density of tactile sensor elements, sampling rate, and resolution of kinesthetic sense all must be high [3], resulting in a high volume of

---

* This chapter is a modified version of a previously published paper Tomoyuki Noda, Takahiro Miyashita, Hiroshi Ishiguro, Norihiro Hagita, *Super-Flexible Skin Sensors Embedded on the Whole Body, Self-Organizing Based on Haptic Interactions*, Robotics Science and Systems Proceedings (RSS 2008), 2008, edited to be comprehensive and fit the context of this book.

tactile information that must be processed. To do so, the following three problems must be solved. First, there is the problem of reduced system robustness due to an increased number of possible failing components. The second is the high cost of data processing. The third is the administration of the sensors' configuration.

The previous study of [6] proposed highly dense distributed skin sensor processing based on interconnecting a self-organized sensor network. Spatiotemporal calculation in each node with spatially seamless tactile information gathered from adjacent nodes enabled haptic interaction features to be extracted, solving the first and second challenges. For instance, an edge detection method is applied to extract features of haptic interaction within the local sensor, yielding efficient data compression. This type of distributed processing requires that the configuration of tactile sensor position be described in the distributed programs of network nodes, which is the remaining third challenge. Manually describing 3-dimensional tactile sensor positions, changing with robot's postures, is very labor intensive and error prone. Moreover, distributed processing typically requires predefined sensor banks, defining which tactile sensors are used in distributed processing for each network node, also a labor-intensive task.

In [7], we found that interaction scenarios could be successfully classified simple k-nearest neighbors (KNN) using a novel feature space based on cross-correlation between tactile sensors, achieving performance of 60% in a 13-way forced choice task. We also found that many categories of touch interactions can be easily visualized by arranging sensors into a "Somatosensory Map" using Multi-Dimensional Scaling (MDS) [8] applied to this feature space as a similarity measure. These promising results suggest that this feature space can be effectively used for automatic analysis of touch behaviors in more complex tasks.

In this chapter, we propose a method for learning "self-organizing tactile sensors" using the feature space from [7] to solve the remaining third challenge. In the proposed method, a classifier is constructed using CLAss-Featuring Information Compression (CLAFIC) [9], a type of a subspace method, applied to a dataset consisting of the full cross-correlation-based feature space of [7]. Instead of directly using a learned subspace as the input for a classifier, we select the sensor pairs that are most "useful," that is, have the highest relevance for the classifier output, to form a more compact sensor bank to be used as input to the classifier. Since now different sensor nodes are involved in different classifiers, it is possible to distribute processing around the body, which can be implemented as in-network processing on the self-organizing sensor network [6]. We call each learned sensor bank a "self-organizing tactile sensor." We found that classifiers based on self-organizing tactile sensors could classify interaction scenarios with an accuracy of up to 80% in a 15-way forced choice task, a significant improvement over prior work. The learned subspaces can also be visualized in a "Somatosensory Map," showing sensor point distribution in a 2D plane.

The rest of this chapter is organized as follows. In Section 5.3.2, we describe related work, and contrast our study with others. Section 5.3.3 describes the basic idea, and then details the proposed method for self-organizing tactile sensors. Section 5.3.4 describes experiments dealing with human–robot haptic interaction used to construct a haptic interaction database. In Section 5.3.5, using the database, the performance of the proposed method is shown. Also, the learned subspaces are visualized in the Somatosensory 2D Map. Section 5.3.6 discusses the results, and Section 5.3.7 concludes.

5.3.2 Background and Comparisons

Prior work on studies of robots with tactile sensors has tended to focus on the development of the physical sensors and transmitting sensor data. For instance, [10] proposes a Large-Scale Integration (LSI) technique for processing data from tactile sensors. Iwata et al. [4] demonstrated physical interaction with users via a skin equipped with 6-axis-kinesthetic sensors. Pan et al. [11] and Inaba et al. [2] described tactile sensors using electrically conductive fabric and strings as a whole-body distributed tactile sensor for humanoid robots. And Shinoda et al. [12] proposed a wireless system for transmitting tactile information by burying wireless sensors under the robot "skin." Thus far, this research has been mainly limited to the problems of collecting tactile information, and solving the necessary wiring and physical implementation problems.

Compared to other sensory modalities such as vision and audio, relatively little prior work has been done on processing haptic interaction from incoming tactile sensor signals. Miyashita et al. [1] estimated user position and posture in interaction using whole-body distributed tactile sensors. Naya et al. [3] classified haptic user interaction based on output from tactile sensors covering a robot pet. Francois et al. [13] also classify different two touch styles, namely, "strong" and "gentle." Though the above research classifies human–robot interaction using tactile sensors, these and other prior studies have not to our knowledge been successful in classifying several haptic interactions while robots are interacting with users.

Pierce and Kuipers [14] proposed self-organizing techniques for building a "cognitive map," which represents knowledge of the body corresponding to physical position of sensors. This map shows the position of each sensor installed on the surface of a robot. However, this method will not work out for a robot having a high degree of freedom and soft skin because the positions of the tactile sensors in 3-d "world" coordinates dynamically change during an interaction. Rather than constructing spatial maps to acquire physical sensor positions, our objective is to use, interpret, and visualize underlying haptic interaction features.

Kuniyoshi et al. [15] proposed a method for learning a "Somatosensory Map," showing the topographic relationship of correlations between incoming signals from tactile sensors distributed on the whole-body surface of a simulated baby. In their somatosensory map, highly correlated sensor points are plotted on a 2D plane close to each other. As the result, the map showed the structure of robot body parts rather than the physical sensor positions as in Pierce and Kuipers [14].

In this paper, our goal is similar to that of [15], so that highly correlated sensor points will be located close to each other, and thus we retain the name "Somatosensory Map" [15]. However, we use a different technique to acquire the map, and use real-world human–robot interaction rather than a simulated baby. Moreover, we attempt to classify haptic interactions using correlations between incoming signals from tactile sensors distributed on the whole-body surface.

5.3.3 Self-Organizing Tactile Sensor Method to Decide Sensor Boundaries

5.3.3.1 Basic Idea

Suppose that N tactile sensors are implemented on a robot, and that the i-th tactile sensor stream during one human–robot interaction is called S_i ($i = 1, \ldots, N$), where S_i is a vector of the n time-step sampling result of the sensor outputs ($S_i = \{(S_i)_1, \ldots, (S_i)_n\}$, $(S_i)_t \in \mathbf{R}$). Features need to be extracted from this time series of the data stream. However, less work has been done on processing tactile features than on audio or vision. In conventional works [3,13], since the data are high dimensional, summary statistics, such as mean, standard error, minimum, max, and coefficients of fast Fourier transformation, are computed from one sensor or all of the sensors to be used as features. A feature space defined from one sensor will not be enough when several sensors are activated by touches, for example, distinguishing a finger tap from a hand tap or a tickle. On the other hand, the feature space computed from combining all the sensors is less robust, since the features could be drastically changed if, for example, one sensor is broken. The feature space defined from several sensors, at least from two sensors, could be an effective happy medium.

We proposed a feature space using cross-correlations computed from sensor pairs, satisfying the above condition, in [7]. The cross-correlation is one important statistic in the human tactile system—Dince et al. [16] reported that the discrimination ability of the two-point stimulus is improved when correlated stimulus is added continually to two closely separated points of a human finger. In fact, the visualization results of our feature space, that is, the Somatosensory Map, show characteristics of many categories of touch interactions [7] by arranging sensor points. Figure 5.19 is the Somatosensory Map made from a 2 min interaction between a human and a robot. This presentation simplifies the haptic interaction; for example, distinctive sensor point clusters of both arms' sensors are the result of the subject touching the robot's arms at the same time.

The cross-correlation-based feature space has another advantage: the choice of sensors to be included in a sensor bank for a certain computation can be decided via the elements used in the classifier's selected subspace. Often, the useful subspace is composed of a combination of sensors located in close spatial proximity, since distant sensors usually have low correlated signals and thus have less mutual information than adjacent sensors.

5.3.3.2 Feature Space

A feature vector **a** is computed from the cross-correlation matrix defined by Equation (5.2) as follows:

$$\mathbf{a} = \left(\underbrace{R_{(1,2)}, ..., R_{(1,N)}}_{N-1}, \underbrace{R_{(2,3)}, ..., R_{(2,N)}}_{N-2}, ..., \underbrace{R_{(N-1,N)}}_{1} \right)^{t} \tag{5.1}$$

where $R_{(i,j)}$ is the cross-correlation matrix element at (i, j) between N sensors, that is,

$$R_{ij}\left(S_i, S_j\right) = \frac{C_{ij}}{\sqrt{C_{ii}C_{jj}}} \quad \left(-1 \le R_{ij} \le 1\right) \tag{5.2}$$

where

$$C_{ij} = \sum_{t=1}^{n} \left(\left(S_i\right)_t - \bar{S}_i \right)\left(\left(S_j\right)_t - \bar{S}_j \right) \tag{5.3}$$

is cross variation of (i, j), and $\bar{S}_i$ is the average of time series data S_i.

5.3.3.3 Overview

We construct a classifier for detecting haptic interaction between robot and human that uses the CLAFIC method [9], a type of subspace method. This method represents each class as eigenpairs computed from a training dataset. The subspace method originated from an idea of Watanabe et al. [9] that, as the feature space grows, the dataset will converge to a limited small subspace. The CLAFIC method approximates this subspace with eigenpairs. Since the feature space defined in III-B is also a high-dimensional space of $O(N^2)$, feature vectors of a dataset should also be restricted mostly to a limited feature subspace.

In our proposed method, a dimension reduction of the extracted subspace is additionally applied by selecting base vectors having large inner product values. The classifier output is then calculated from a selected subspace consisting of only a few base vectors chosen from cross-correlation elements of the coefficient R_{ij} computed from all sensor pairs. Hence, this subspace and feature selection results in useful subsets of sensors that can be used to distribute processing. Since these subsets are found automatically, we call this method finding "self-organizing tactile sensors."

5.3.3.4 CLAFIC Method

Figure 5.13 shows the overview of classification. At first, a dataset $\mathbf{X}_k$ is prepared from p_k feature vectors $(\mathbf{a}_k)$ computed from a time series of sensor streams labeled as a class $k \left(\stackrel{def}{\equiv} \omega_k \right)$, where p_k is the number of training dataset. Thus

$$\mathbf{X}_k = \left\{ a_{k1}, ..., a_{kp_k} \right\} \tag{5.4}$$

The approximated subspace of CLAFIC method begins by performing singular value decomposition (SVD):

$$\mathbf{X}_k = \mathbf{U} D_\lambda \mathbf{V}^t \tag{5.5}$$

where the columns of $\mathbf{U}$ are left singular vectors; D_λ has singular values and is diagonal; and $\mathbf{V}^t$ has rows that are right singular vectors. We perform a forward feature selection of singular vectors to perform classification using a low-dimensional subspace. First, we arrange the vectors obtained from the SVD performed only on data from class k in decreasing order of the $((d_k)_1, (d_k)_2, ..., (d_k)_{smallest})$ singular values, and compute a discrimination function DF_k of class k, which takes input an unknown vector $\mathbf{x}$, and computes

$$DF_k(\mathbf{x}) = \sum_{j=1}^{m} \left(\mathbf{x}^t \mathbf{u}_{kj} \right)^2 \tag{5.6}$$

using only the first m vectors, where $\mathbf{u}_{kj}$ is a left singular vector derived only from data in class k. We choose m by starting at $m = 1$ and increasing m until

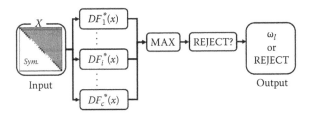

FIGURE 5.13
Overview of the classification. (a) Robovie-IIF (ATR), (b) Tactile sensor network consisting of an RS422 bus network via which nodes are connected to a host PC

the cumulative contribution ratio in eq. 6 exceeds a threshold value C_1. The output of the discrimination function corresponds to the square of the length of an unknown vector orthographically projected onto the low-dimensional subspace. The classifier outputs the class name which has maximum *DF* output.

$$\max_{k=1,\dots,c} \left\{ DF_k(\mathbf{x}) \right\} = DF_l(\mathbf{x}) \quad \mathbf{x} \in \omega_l \tag{5.7}$$

Using the fidelity value τ as Watanabe et al. propose [9], the unknown is vector classified as "unknown" class if the maximum *DF* is not significantly different from the second maximum *DF*, that is, if

$$\frac{DF_l(\mathbf{x})}{\max_{k \neq 1}\left(DF_k(\mathbf{x})\right)} > \frac{1}{\tau} \tag{5.8}$$

evaluates to "false," the classification is "unknown."

5.3.3.5 Learning Sensor Banks

A sensor bank for a classification task is decided by selecting a useful subspace that has high relevance for the classifier. As Equation (5.6) shows, the output of the classifier is composed of inner products. Considering that the $\mathbf{u}_{ij}$ elements are weights for the unknown vector elements, if a p-th element $\{\mathbf{u}_{ij}\}_p$ is close to 0, the element $\{\mathbf{x}\}_p$ could be ignored. In computer vision, Ishiguro et al. [17] has proposed this type of idea, describing it as a form of "attention control." Let $\tilde{\mathbf{u}}_{ij}$ be an approximated subspace ignoring the all elements close to 0, where all $\{\mathbf{u}_{ij}\}_p$ smaller than C2 are simply replaced by 0. Now we construct an approximated discrimination function $DF_k^*(\mathbf{x})$ as follows.

$$DF_k^*(\mathbf{x}) = \sum_{j=1}^{m} \left(\mathbf{x}^t \mathbf{u}_{kj}^*\right)^2 \tag{5.9}$$

where $\mathbf{u}_{kj}^*$ are assumed to be nearly orthogonal, and are normalized $\left(\mathbf{u}_{kj}^* = \tilde{\mathbf{u}}_{kj} / \left\| \tilde{\mathbf{u}}_{kj} \right\| \right)$.

The approximated discrimination function $DF_k^*(\mathbf{x})$ is also composed of inner products of $\mathbf{u}_{kj}^*$ and $\mathbf{x}$; however, now we need to know only the q-th elements $(\{\mathbf{x}\}_q)$, where

$$\left\{ q : \left\{ \mathbf{u}_{kj}^* \right\}_q \neq 0 \right\} \tag{5.10}$$

From the definition of feature space in Equation (5.1), only elements $\mathbf{R}_{(r_q, s_q)}$ need to be computed, where (r_q, s_q) is a sensor pair needed to compute the

q-th element $\{x\}_q = \mathbf{R}_{(r_q, s_q)}$. These sensor pairs define whether the sensor is used or not used in the sensor bank, facilitating distributed processing, since each classifier only needs a subset of sensors.

5.3.3.6 Somatosensory Map

To visualize feature vectors and sensor banks, we define dissimilarity as d_{ij} converted from the coefficient of the cross-correlation matrix R_{ij} with the following equation:

$$d_{ij}\left(R_{ij}\right) = -\log\left(\left|R_{ij}\right|\right). \quad (0 \le d_{ij} \le \infty) \tag{5.11}$$

This dissimilarity definition defines a "distance" between (i, j) sensors; that is, higher correlated (or negatively correlated) sensor pairs have smaller dissimilarity. (Note that it does not satisfy all properties of a true distance notion.) Since the self-correlation coefficient is always 1, the dissimilarity with itself is always 0, that is,

$$d_{ij} = 0. \; (S_i \ne \text{constant}) \tag{5.12}$$

In the Somatosensory Map, the N sensor points are arranged into a 2D map using MDS [8] based on the dissimilarity definition of Equation (5.11). This 2D map can be used to visualize a vector of the cross-correlation feature space; for example, Figure 5.19 is the Somatosensory Map plotted using a feature vector during a human–robot interaction [7]. In Section 5.3.5, we apply this method to $\mathbf{u}_{ij}$ to interpret experimental results.

5.3.4 Experiments

5.3.4.1 Hardware

The hardware on which we are testing our proposed technique is detailed below. Figure 5.14 shows an outline of the hardware for the experiments described in this section. Figure 5.14a shows the communication robot Robovie-IIF [20], provided with high-density, soft-tactile sensors and a sensor network consisting of a RS422 bus network via which nodes are connected to a host PC (Figure 5.14b). Figure 5.15 shows the structure and materials of the skin sensors installed on the Robovie-IIF surface, and Figure 5.16 is the location of embedded piezofilms.

Two hundred seventy four Piezofilms (3 cm × 3 cm, or 5 cm × 5 cm) are embedded in soft silicone rubber. Sampling time for the 16-bit A/D converter is set to 100 Hz, and tactile sensor outputs are read to the host PC of Robovie-IIF at every sampling. Using this hardware, we conducted two experiments detailed below and shown in Figures 5.17 and 5.18.

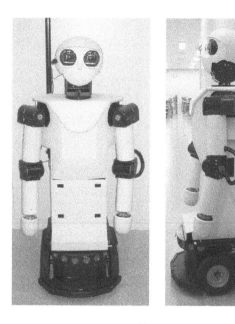

(a)

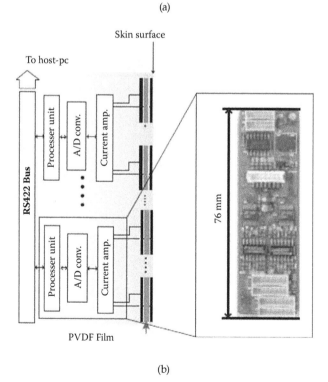

(b)

FIGURE 5.14
Overview of our tactile sensor system.

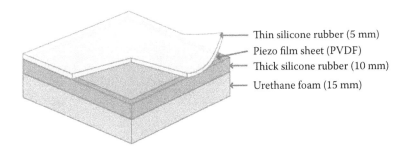

FIGURE 5.15
Architecture of skin sensor devices.

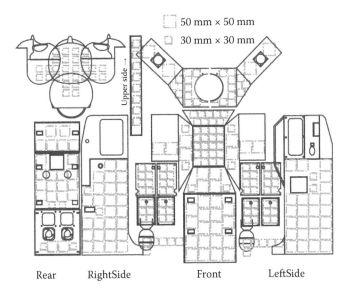

FIGURE 5.16
Position of the skin sensors (PVDF films) on the deployed surface of the Robovie-IIF. (a) Pat the head. (b) Hug. (c) Touch the robot.

5.3.4.2 Field Experiment

Our research group conducted several events in the Osaka Science Museum with socially interactive robots [19]. The first experiment was a field experiment during the event named "Let's play with Robovie." In the event, the Robovie-IIF was displayed in the Osaka Science Museum during May 2005. We asked visitors to play with the Robovie-IIF with the goal of investigating what kind of haptic interaction can be realized between humans and the Robovie-IIF. Figures 5.17 (a) to (c) show the three haptic interactions such as (a) Patting on the head, (b) Hug, and (c) Touch the robot body observed in this experiment. From these observations, we designed a "haptic interaction scenario" for a second experiment to encourage human–robot interaction in

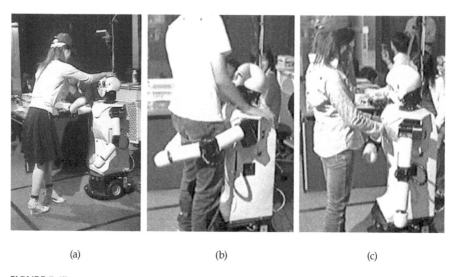

(a)   (b)   (c)

FIGURE 5.17
Observed haptic interactions between the robot and the visitors at Osaka Science Museum, May 2005. (a) "I wish you'd pat me on the head." (b) "Tickle me." (c) "Give me a hug."

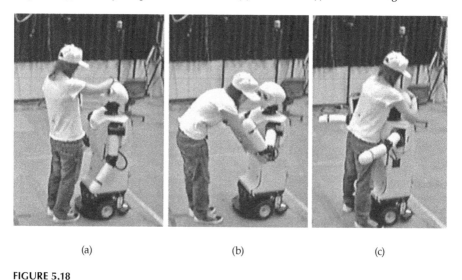

(a)   (b)   (c)

FIGURE 5.18
The observed subject's behaviors in each step of the scenario during the experiment for database construction.

which we expected subjects to touch Robovie-IIF (detailed below). Table 5.2 shows the interaction scenario stages. Each stage in the scenario consists of the control rule that the Robovie-IIF tries to sustain the interest of a subject to keep the interaction going and proceeds to the next stage after finishing each interaction. (See "approaching a person" of Table 5.2.)

TABLE 5.2

Scenario-Based Definition of the Interaction Classes

Class Name (Step)	Approaching a Person	m ($C_1 = 0.15$)
*class*1	"Hello."	4
*class*2	"Let's shake hands."	4
*class*3	"Nice to meet you."	5
*class*4	"What's your name?"	5
*class*5	"Where are you from?"	5
*class*6	"Let's play!"	5
*class*7	"Do you think I'm cute?"	4
*class*8	"I wish you'd pat me on the head."	3
*class*9	"Whee!"	6
*class*10	"I want to play more."	5
*class*11	"Tickle me."	7
*class*12	"That tickles!"	3
*class*13	"Thanks."	4
*class*14	"Give me a hug."	7
*class*15	"Bye-bye!"	4

5.3.4.3 Construction of the Haptic Interaction Database

In the second experiment, using a "Wizard of OZ" method [18] based on the scenario-based rules described in Section 5.3.4.2, Robovie-IIF was controlled by an experimenter with several monitoring displays. We expected the robot to be touched by subjects during these interactions. We constructed a scenario-based interaction database, which includes the monitoring videos, all of the tactile sensor signals, the command signals sent to control the robot, with all data time-stamped using a common clock.

Each subject was asked to interact with Robovie-IIF in 3 trials separated by 10 minutes each. There were a total of 48 subjects, 24 males and 24 females, all of them college students. Each trial took around 5 minutes and was set up with the same condition except for the subject's position at the start of the scenario. These positions were each 2 m away from Robovie-IIF, at 45° to the right, 45° to the left, and 0° (where 0° is defined as in front of the robot). We asked the subjects to simply play with Robovie-IIF (which has a child-like voice and uses other cues to encourage humans to treat it as a child), and explained to them before each trial the following rules: (1) the subjects can touch the whole body of Robovie-IIF, (2) they are required to listen carefully to what Robovie-IIF is saying, and (3) they are required to be close to the robot in order to turn it on by touching it at the start of each trial.

Figure 5.18a–c show the observed haptic interactions in the experiment, such as (a) "I wish you'd pat me on the head" of *class*8, (b) "Tickle me" of *class*11, and (c) "Give me a hug" of *class*14. Excluding approximately 24 cases in which there were technical difficulties, approximately 120 cases of data were acquired to form a "haptic interaction database" of data collected from real interaction scenarios. Segments of the tactile sensor data are automatically clipped and labeled using the time stamps for when each scenario stage (as defined in Table 5.2) begins and ends. Thus, unlike previous work in which interaction segments were hand-labeled by an experimenter, we do not perform any manual coding of the data.

5.3.5 Results

To emphasize the tactile sensor data when subjects touch the robot, we prepared from the output of tactile sensors as shown in Equation (5.13) since our database includes lots of information caused by the robot's movements.

$$
\hat{S} = \begin{cases} S_i, \left(\left| S_i - \bar{S}_i \right| > \sigma_i \right) \\ \bar{S}_i, \left(\left| S_i - \bar{S}_i \right| \le \sigma_i \right) \end{cases} \tag{5.13}
$$

where S_i is the i-th sensor output, the average of time series data of S_i is $\bar{S}_i$, and the standard deviation of S_i is σ_i. If the absolute difference between S_i and $\bar{S}_i$ is smaller than the standard variation σ_i, S_i is replaced by the average $\bar{S}_i$. In this section, all of the results are from $\hat{S}_i$.

Figure 5.19 shows the 2D Somatosensory Map obtained from a cross-correlation matrix of each tactile sensor during an interaction in a field experiment between the Robovie-IIF and a subject (Figure 5.17c). Figures 5.20 and 5.21 show the results of Leave-One-Out cross-validation tests for evaluation of the classifier using the K-nearest neighbor (KNN) method ($k = 3$) and the currently proposed method, respectively, on the whole dataset. Figures 5.21 through 5.24 show the results of choosing different values of C1 and C2 using the proposed method. In these figures, *class k* ($k = 1, \ldots, 15$) corresponds to the classes defined in Table 5.2. For example, the dataset labeled *class2* consists of tactile sensor data collected between the time when the start command of *class2* ("Let's shake hands") was sent to the Robovie-IIF and the time when the start command of *class3* ("Nice to meet you") was sent. (The datasets include some cases in which subjects did not deliver the expected interaction.) As can be seen in Figure 5.20, the KNN method achieved classification of over 60% for many haptic interactions such as *class2*, *class6*, *class7*, *class8*, *class11*, and *class15*, using only the correlation patterns of all tactile sensors.

Figure 5.21 shows the correct recognition rates for the proposed method, while Figure 5.22 shows the "false alarm" rate, computed for each class as the number of times an example was incorrectly classified as belonging to that class, divided by the number of examples that actually belong to that

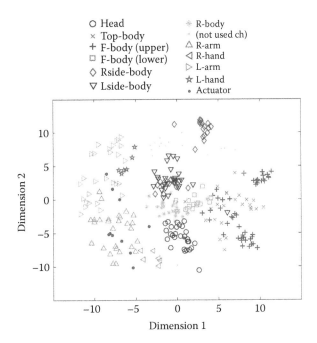

FIGURE 5.19
This map [7] is the 2D Somatosensory Map obtained from the haptic interaction shown in Figure 5.17(c) in the field experiment. The time window used to compute the distance matrix, $\{dij\}$, is 2 min of the whole interaction between the robot and the subject. The two clusters of the "F-body (upper)" sensors' distribution, around the locations of $(Dimension1, Dimension2) = (7, -6)$ and (11,4), can be interpreted as the result of that she touched the two areas of the upper body with her left/right hands. This result suggests that cross-correlations contain touch features.

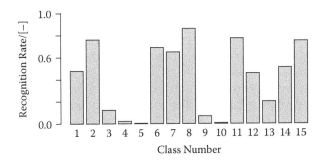

FIGURE 5.20
Recognition Rate of the classifier using the k-nearest neighbor method in 15-way forced choice task ($k = 3$ is the same condition as [7], but with 2 classes added to the choice. [7] was in 13-way forced choice task.).

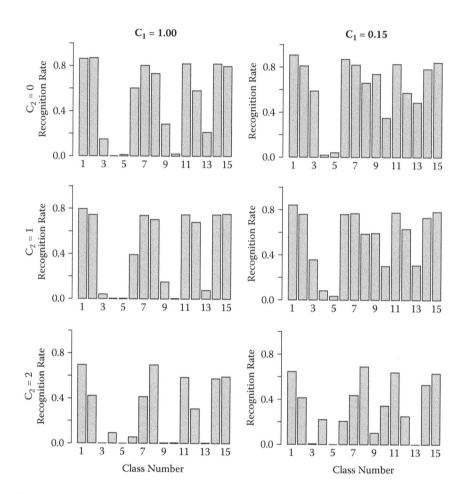

FIGURE 5.21
Recognition rate.

class (thus these numbers can be greater than 1). For these experiments, the fidelity value was experimentally fixed to be 0:95, which did not change the recognition rate but improved the false alarm rate. Each figure has 6 conditions that are in the set $\{(C_1,C_2) : C1 = 1, 0.15\ C_2 = 0,1, 2\}$. The number of orthogonal base vectors (that are left singular vectors, $\mathbf{u}_{ij}$) is decided by the parameter C_1, shown in Table 5.2. The parameter C_2 determines the reduction of size of the feature space; for example, the reduced feature spaces in the case $C_2 = 0, 1, 2$ were 0%, around 80–85%, and around 94–96%, respectively.

The proposed method improved classification to 80% for most haptic interactions, including *class*1, *class*2, *class*6, *class*7, *class*9, *class*11, *class*14, and *class*15 when using $(C_1,C_2) = (0,15, 0)$. This performance was almost the same as for $(C_1,C_2) = (0.15, 1)$, which used only 15% of the feature space. When the feature space size is reduced to 5% in the condition $(C_1,C_2) = (0.15, 2)$, performance is

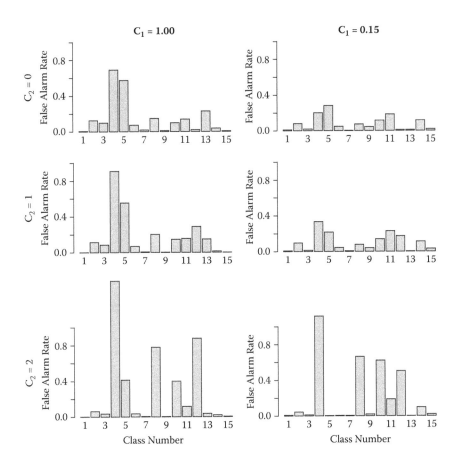

FIGURE 5.22
False alarm rate.

still as high as 60% for haptic interactions of *class*1, *class*8, *class*11, and *class*15 (note that random, "by chance" performance is less than 7%).

Figure 5.23 shows the result of the feature space reduction. Elements of vector $\mathbf{u}_{81}$ are arrayed onto a matrix of the same size as a cross-correlation matrix, and large weighted cross-correlation elements are visualized with darker (more black) colors. Adjacent tactile sensors usually have closer numbers, and are displayed as square line boxes corresponding to their part names. As expected, the useful feature spaces are composed mostly of adjacent sensor pairs. The boxes shown, which include several highly weighted elements, are the result of self-organizing results corresponding to the boundaries of tactile sensors. These self-organizing results are also shown in Figure 5.24a and 5.24b, which shows the arrangement of sensors in a 2D Somatosensory Map using $\mathbf{u}_{(8)1}$ and $\mathbf{u}_{(11)1}$. Closer sensor pairs have larger weight in their cross-correlation element. Figure 5.24a shows that head sensors, probably touched in the "pat me" interaction, are clustered apart from

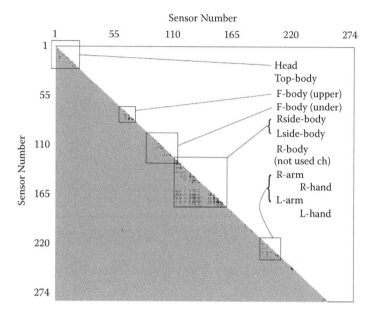

FIGURE 5.23
Color image matrix reconstructed from orthogonal vector **u**81, highlighting the large weighted elements of the cross-correlation. (a) Somatosensory map made from u(8)1 (class 8). (b) Somatosensory map made from u(11)1 (class 11).

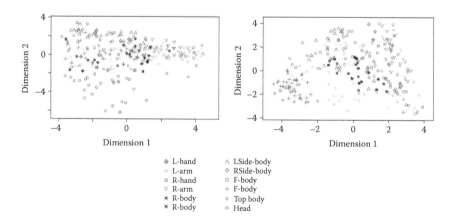

FIGURE 5.24
Somatosensory map can also visualize a **u**kj as a sensor point distribution in a 2D plane, since **u**kj has the same dimension as the feature vector **a**. In these two maps, closer sensor pairs have larger weight in their cross-correlation element.

other sensors. Figure 5.24b also shows clustering of the front side of the body (F-body) and of the left and right side of the body (LSidebody and RSidebody), which are often touched together in the "tickle" interaction. Note that these results do not make use of any knowledge of the spatial position of sensors but only use sensor streams from the whole robot body.

5.3.6 Discussion

Comparing our Somatosensory Map with previous work [14,15], we found that haptic interactions form clusters in the map that often can be grouped by body part. Using this representation for human–robot interaction, we achieved good classification results for those interaction categories in which there was some human touching. In previous work for classification of haptic interaction [3,13], the datasets for learning classifiers were hand-labeled by the experimenter. In our case, the database is self-labeled during scenario-based interactions. The label of each data point is based on what the current designed scenario is, rather than given *post hoc* by the experimenter asking subjects to touch the robot. We assume this is a more natural and practical database construction.

The Somatosensory Map shows large weighted elements mainly between spatially localized sensors. This is consistent with the idea of Watanabe et al. that, as the feature space grows, datasets converge to limited subspaces. Additionally, the learned subspace was composed mainly of adjacent sensor pairs in the tactile system, as seen in Figure 5.23. Thus, the CLAFIC method is able to achieve higher classification performance even when using smaller subspaces of 15% size. In fact, a performance of 80% classification was achieved, improving over the KNN method of 60%, despite a much more challenging evaluation than in previous work. Instead of using static objects consisting of tactile sensors, we constructed the database from real human robot interactions. Since the robot has malleable tactile sensors embedded under soft skin, and the robot is moving during the experiment, it is possible that the results come from the classification of self-sensations provided by self-movements. On the other hand, it seems that the information from subjects' touch improved the classification performance, because the recognition rate were low in the classes in which subject were less likely to touch the robot. We believe that further improvements should be considered, for example by incorporating proprioception in the tactile system, since a robot would be in a motion when a human touches it.

5.3.7 Conclusions

In conclusion, the proposed method was found to be efficient with the classification of real human–robot interactions, and was able to be implemented as distributed in-network processing.

In this chapter, we describe a haptic interaction classification method using cross-correlation matrix features, and propose a self-organizing technique to define a bank of sensors to be used in distributed processing of each class. The cross-validation tests results in recognition of 80% for those interactions in which we expect subjects to touch the robot, using only 15% of the feature subspace. The Somatosensory Map visualization shows that the selected feature space was composed mainly of spatially adjacent sensor pairs. These promising results suggest that our proposed method may be useful for automatic analysis of touch behaviors in more complex future tasks.

Acknowledgments

Valuable comments from Dr. Ian Fasel improved the presentation quality of this chapter. The research in Section 5.3.4 was supported in part by Japan's Ministry of Internal Affairs and Communications. The analysis of the experimental data in Section 5.3.5 was conducted as part of the JST ERATO Asada Project.

References

1. T. Miyashita, T. Tajika, K. Shinozawa, H. Ishiguro, K. Kogure, and N. Hagita, Human position and posture detection based on tactile information of the whole body, In *Proc. of IEEE/RSJ 2004 International Conference on Intelligent Robots and Systems Workshop (IROS'04 WS)*, Sep. 2004.
2. M. Inaba, Y. Hoshino, K. Nagasaka, T. Ninomiya, S. Kagami, and H. Inoue, A full-body tactile sensor suit using electrically conductive fabric and strings, *Proc. 1996 IEEE/RSJ International Conference on Intelligent Robots and Systems (IROS 96)*, Vol. 2, pp. 450–457, 1996.
3. F. Naya, J. Yamato, and K. Shinozawa, Recognizing human touching behaviors using a haptic interface for a Pet-robot, *Proc. 1999 IEEE International Conference on Systems, Man, and Cybernetics (SMC'99)*, pp. II-1030–1034.
4. H. Iwata, H. Hoshino, T. Morita, and S. Sugano, Force detectable surface covers for humanoid robots, *Proc. 2001 IEEE/ASME International Conference on Advanced Intelligent Mechatronics (AIM'01)*, pp. 1205–1210, 2001.
5. H. Iwata and S. Sugano, Whole-body covering tactile interface for human robot coordination, *Proceedings—IEEE International Conference on Robotics and Automation*, Vol. 4, pp. 3818–3824, 2002.
6. T. Noda, T. Miyashita, H. Ishiguro, K. Kogure, and N. Hagita, Detecting feature of haptic interaction based on distributed tactile sensor network on whole body, *Journal of Robotics and Mechatronics*, Vol. 19, No. 1, pp. 42–51, 2007.
7. T. Noda, T. Miyashita, H. Ishiguro, and N. Hagita, Map acquisition and classification of haptic interaction using cross correlation between distributed

tactile sensors on the whole body surface, *Proceedings of IEEE/RSJ International Conference on Intelligent Robots and Systems (IROS 07)*, 2007.

8. W. Torgerson, Multidimensional scaling: I. Theory and method, *Psychometrika*, 17, pp. 401–419, 1952.

9. S. Watanabe and N. Pakvasa, Subspace method in pattern recognition, *Proc. 1st Int. J. Conf on Pattern Recognition*, Washington DC, pp. 2–32, Feb. 1973.

10. A. Iwashita and M. Shimojo, Development of a mixed signal LSI for tactile data processing, in *Proc. of IEEE Int. Conf. on Systems, Man and Cybernetics 2004 (SMC2004)*, 2004.

11. Z. Pan, H. Cui, and Z. Zhu, A flexible full-body tactile sensor of low cost and minimal connections, *Proc. 2003 IEEE International Conference on Systems, Man, and Cybernetics (SMC'03)*, Vol. 3, pp. 2368–2373, 2003.

12. H. Shinoda, N. Asamura, T. Yuasa, M. Hakozaki, X. Wang, H. Itai, Y. Makino, and A. Okada, Two-dimensional communication technology inspired by robot skin, *Proc. IEEE TExCRA 2004 (Technical Exhibition Based Conf. on Robotics and Automation)*, pp. 99–100, 2004.

13. D. Francois, D. Polani, and K. Dautenhahn On-line behaviour classification and adaptation to human-robot interaction styles, poster session of HRI (2007), HRI, 2007.

14. D. Pierce and B. Kuipers, Map learning with uninterpreted sensors and effectors, *Artificial Intelligence*, Vol. 92, 1997.

15. Y. Kuniyoshi, Y. Yorozu, Y. Ohmura, K. Terada, T. Otani, A. Nagakubo, and T. Yamamoto, From Humanoid Embodiment to Theory of Mind, July 7–11, 2003, revised papers, pp. 202–218.

16. Hubert R. Dince et al., Improving human haptic performance in normal and impaired human populations through unattended activation-based learning, *ACM Transactions on Applied Perception*, Vol. 2, No. 2, pp. 71–88, April 2005,

17. H. Ishiguro, M. Kamiharako, and T. Ishida, State space construction by attention control, *International Joint Conference on Artificial Intelligence (IJCAI-99)*, pp. 1131–1137, 1999.

18. N. Dahlback, A. Jonsson, and L. Ahrenberg, Wizard of Oz studies—why and how, *Proc. of the International Workshop on Intelligent User Interfaces*, pp. 193–200, 1993.

19. M. Shiomi, T. Kanda, H. Ishiguro, and Norihiro Hagita, Interactive humanoid robots for a science museum, *IEEE Intelligent Systems*, Vol. 22, No. 2, pp. 25–32, Mar/Apr, 2007.

20. T. Miyashita, T. Tajika, H. Ishiguro, K. Kogure, and N. Hagita, Haptic communication between humans and robots, in *Proc. of 12th International Symposium of Robotics Research*, CD-ROM, San Francisco, CA, Oct. 2005.

5.4 Integrating Passive RFID tag and Person Tracking for Social Interaction in Daily Life

Kenta Nohara, Tajika Taichi, Masahiro Shiomi, Takayuki Kanda, Hiroshi Ishiguro, and Norihiro Hagita

ABSTRACT

This article reports a method in which a communicative robot simultaneously interacts with two persons and identifies them with passive-type RFID and floor sensors. The difficulty emanates from the association of these inputs. A passive-type RFID reader provides very accurate person identification. However, the robot cannot specify who actually used the RFID if surrounded by many people. A floor sensor provides robust tracking of people's positions, but it does not identify any person by itself. Thus, combining these sensors into one device that provides robust tracking of position with accurate person identification would be very beneficial. Toward this problem, in our key idea, the robot considers multiple hypotheses about the interpretation of sensor information and gradually improves their accuracy through interaction with the people. The multimodal inputs from sensors are integrated with a Bayesian network that allows us to exploit people's behavior patterns as parameters to estimate the accuracy of our hypotheses. Our method also considers the vagueness of current hypotheses when choosing appropriate interactive behaviors. When the robot has reservations about the situation, it prefers nonconclusive behavior and tries to solve the ambiguity by performing verification behaviors. Once the robot is certain, it prefers conclusive behavior. The developed system was tested in a real daily environment, a shopping center, where we gathered interaction data between the robot and multiple people as teaching data for the Bayesian network. The experimental results revealed that the system successfully identified 79.2% of the visitors. In particular, the hypothesis-based system for behavior control improved the successful rate by 16.7%.

5.4.1 Introduction[*]

The development of robots is entering a new level, where the focus is being placed on interaction with people in daily environments. In fact, several communication robots have already been developed for everyday environments visited by many people, including an elementary school [1], museums [2,3], an Expo [4], and a train station [5]. Thus, the concept of communication robots continues to rapidly evolve. In the future, humanoid robots in public or commercial spaces might provide information on route guidance or nearby shops (Figure 5.25).

The importance of the identification of a person is increasing with advancement of the possibilities of robot services. For instance, past works report the efficiency of behavior based on such personal information as the robot's calling

---

[*] This chapter is a modified version of a previously published paper Kenta Nohara, Taichi Tajika, Masahiro Shiomi, Takayuki Kanda, Hiroshi Ishiguro, and Norihiro Hagita, *Integrating Passive RFID Tag and Person Tracking for Social Interaction in Daily Life*, Proceedings of the 17th IEEE International Symposium on Robot and Human Interactive Communication (RO-MAN2008), pp. 545–552, 2008, edited to be comprehensive and fit with context of this book.

FIGURE 5.25
Service robot in a shopping center.

visitors by name [1,2]. Moreover, a robot can choose conversational topics based on information about visitors. There are two main methods for person identification: one is contactless identification, such as using an active-type RFID tag [1,2,6] with face recognition [7]. The other method is contact identification, such as using a passive-type RFID tag [8] with biometrics [9].

We chose a passive-type RFID tag contact identification technology for person identification to focus on the difficulties of adapting contactless identification techniques to a robot in real daily environments. Many kinds of noisy information surface in such environments, including changes in lighting conditions and the existence of multiple people. In fact, past situations report that many visitors often surround one robot [1,2]. Moreover, robot movement that depends on interactive behavior will create difficulties for sensor processing. Moreover, a contactless identification system has a privacy problem: the system can identify a person without permission.

However, a passive-type RFID reader cannot specify who actually used the RFID if multiple people are around the robot, although it provides very accurate person identification. Therefore, we proposed a method with which a communication robot simultaneously interacts with two persons by identifying them with passive-type RFID and floor sensors. The floor sensors provide robust tracking of people's positions [10] without identifying any person. Thus, we thought that we could complement each sensor's weakness by combining them.

We propose a unique method that realizes robust people tracking with accurate person identification and allows a robot to perform hypotheses not only with a sensor layer but also with an interaction layer. In other words, there are two main ideas. The first is that the robot considers multiple hypotheses about the interpretation of sensor information. Multimodal inputs from sensors are integrated with a Bayesian network [11] that enables us to use people's behavior patterns as parameters to estimate the accuracy of hypotheses. Our method also uses the vagueness of current hypotheses to choose appropriate interactive behaviors. The second main idea is that the robot gradually improves the accuracy of hypotheses through interaction with people. For example, when the robot has reservations about the situation, it chooses nonconclusive behavior and tries to resolve the ambiguity by performing such verification behaviors as searching through the faces of the people interacting with the robot and estimating distance relationships with them; once the robot is certain, it prefers conclusive behavior.

In this article, we describe the effects of the proposed method on a communication robot that simultaneously interacts with two persons while identifying them with a passive-type RFID tag reader and floor sensors. We implemented the proposed method on a communication robot that interacts with visitors using personal information in a shopping center. The experimental results revealed the accuracy of the proposed method during ambiguous situations.

5.4.2 Person Tracking and Identification

In this study, we focus on a situation where a robot simultaneously interacts with two or more people and has to identify them to provide information depending on personal information. In such situations, the robot needs to track the positions of people and identify them. In other words, it needs the simultaneous, integrated capabilities of both person tracking and identification. With such functionalities, the robot can continue interacting with a certain identified person when many other people are around it. As shown in Figure 5.25, we anticipate that communication robots will soon start to work in public spaces with many people; thus, it is reasonable to imagine such situations for communication robot services. In this section, we describe the advantages and the details of the proposed method by comparing existing person identification methods.

5.4.2.1 Previous Methods

Table 5.3 compares person tracking and person identification methods. This issue remains unresolved in robotics.

Image processing is a well-studied strategy that uses a robust device that provides both functionalities. Many studies track people and identify them from faces. However, this approach suffers from two difficulties for

TABLE 5.3

Comparison of Person Identification Methods for a Robot to Identify the Target Person from Multiple People

	Accuracy of Person Tracking	Person Identification	Robustness	Weakness
Floor sensor [10]	~20 [cm]	—	Occlusion, lighting	Identification
Laser range finder	~5 [cm]	—	Lighting	Occlusion, identification
Camera (Vision)[7]	~5 [cm]	Contactless	—	Occlusion, lighting
Active type RFID tag [6]	~3 [m]	Contactless	Lighting	Reflection, crowdedness
Passive type RFID tag [8]	—	Contact	Occlusion, lighting	Person tracking
Proposed method [This paper]	~20 [cm]	Contact	Occlusion, lighting, ambiguity of situations	—

achieving our purpose. One is a weakness in real environments, such as shopping centers and public spaces, because occlusions frequently arise due to the presence of many people. Changes in lighting conditions also arise in such environments. The other problem concerns privacy; person identification with image processing achieves contactless identification. Such functions include the risk of infringing on the privacy of others.

An active-type RFID tag technology enables a robot to easily identify a person using radio signals. With multiple RFID tag readers, a person-tracking function is achieved; however, its accuracy is less than 3 m. Thus, only with this accuracy, the robot cannot approach the target person and continue the interaction. In addition, an active-type RFID tag also raises the same privacy problems as the image processing function.

From these points of view, achieving our purpose is difficult using only one kind of sensor; we need to integrate different kinds of sensors. Thus, another strategy is to combine two or more devices to achieve both functionalities. Many well-established methods track people positions. Floor sensors and laser range finders easily achieve robust tracking functions using condensation algorithms. These devices provide accuracy of less than 20 cm, a reasonable range for a social robot. It can approach them, turn its body toward them, and look in their direction. For person identification, a passive-type RFID tag technology easily enables a robot to identify a person.

5.4.3 Proposed Method

We propose a method that integrates a person identification function provided by a passive RFID and a person tracking function provided by floor

FIGURE 5.26
A scene of a difficult situation in identifying people.

sensors. Both devices provide robust recognition by themselves. We put a passive-type RFID reader on the robot to enable it to identify people who interact with it by RFID tag.

However, the fundamental difficulty concerns how to associate such different inputs. If only one person is near the robot when it detects an RFID tag, associating the closest person's position and RFID information is easy. Figure 5.26 shows an example of the difficulty: if two or more people are around the robot, associating the information from the RFID tags and the position of the people is difficult.

To solve this association problem, the robot must actively behave in social interaction to solve ambiguity caused at the sensor level. The proposed method uniquely integrates three methods: combining floor sensors and a passive-type tag reader, identifying people using multiple hypotheses based on gathering realistic interaction data, and gradually improving the accuracy of the robot's identification hypotheses through interaction with people. In this section, we describe the details of each component of the proposed method.

By integrating these sensors, the system resolves ambiguity in person identification. Several past works successfully resolved such ambiguity based on hypothesis approaches; for example, Fransen et al. developed a robot system [12] that solves ambiguous situations with sensor layers. Inamura et al. developed a humanoid robot to solve ambiguous dialogue situations with interaction layers [13]. Considering these works, we focus on Bayesian networks that allow us to use people's behavior patterns as parameters to estimate the accuracy of hypotheses. Thus, we make hypotheses by integrating floor sensors and a passive-type RFID tag reader with a Bayesian network.

The robot gradually improves the accuracy of its hypotheses through interaction with people when it has reservations about the situation. In such ambiguous situations, the robot prefers nonconclusive behavior and tries to solve the ambiguity by performing verification behavior. For example, the relationship distance between the robot and a person is important information for verifying hypotheses when the robot addresses someone while shaking hands. If the robot detects a change in the relationship distance, it can update the accuracy of the hypothesis. Once the robot is certain, it prefers conclusive behavior. Thus, the robot actively behaves in social interaction to solve ambiguity caused at the sensor level.

5.4.4 System Configuration

5.4.4.1 System Overview

Figure 5.27 shows the entire robot system with our proposed method. Our system simultaneously interacts with two persons while identifying them. The robot's services include conversation using personal information. The developed system consists of a humanoid robot with a passive-type RFID tag reader, floor sensors, and the proposed method. The robot interacts with two people at the same time, based on the hypotheses. The robot gradually improves its hypothesis accuracy through interaction with people. The details of each component are described below.

5.4.4.2 Humanoid Robot

We used Robovie, a humanoid robot for this system. Figure 5.28 shows "Robovie-IIF" [14], an interactive humanoid robot characterized by its human-like physical expressions and its various sensors. This robot is based on Robovie-II [15] and has tactile sensor elements embedded in a soft skin

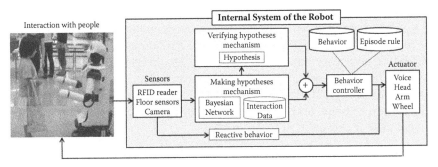

FIGURE 5.27
Robot control system.

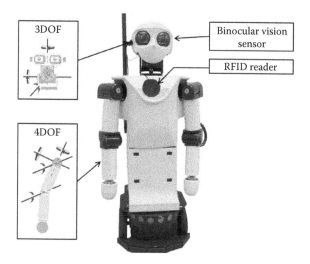

FIGURE 5.28
RovovieII-F.

that covers the entire robot's body. We used humanoid robots because a human-like body is useful to naturally catch people's attention.

The human-like body consists of a body, a head, a pair of eyes, and two arms. When combined, these parts can generate the complex body movements required for communication. It is 120 cm high with a 40 cm diameter, two 4×2 degrees of freedom in its arms, 3 degrees of freedom in its head, and a mobile platform. It can synthesize and produce a voice through a speaker.

The behavior controller executes a behavior based on sensor information and the *Episode rules* that govern the order of execution. The actuator modules then actuate the robot to interact with people. Reactive behaviors include avoidance and gazing at a touched part of its body as well as such patient behavior as playing by itself. These reactive behaviors are controlled based on robot sensor information from tactile, vision, and distance sensors.

5.4.4.2.1 Person Identification Function

To realize a person identification function, we installed a passive-type RFID tag reader on the robot (Figure 5.28). Figure 5.29 shows a passive-type RFID tag and an accessory embedded RFID tag. The height of the accessory is 4 cm. The detection area of the tag reader is less than 15 cm. This RFID tag can record a unique ID, such as "A" and "B." We incorporated such personal information as name and height in the ID of the RFID tags so that the robot can estimate it using the RFID tag reader.

5.4.4.2.2 Face Detection Function

To verify the reliabilities of the hypotheses, the robot needs a function to find the faces of people with whom it is interacting. We modified our previous

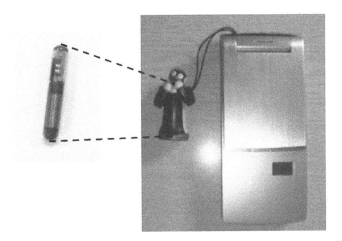

FIGURE 5.29
RFID tag and accessory embedded in RFID tag.

face detection function [16] with height and distance information to allow the robot to estimate the location of the interacting people's face more robustly. The height information is provided by the personal identification function. The distance information is provided by the human tracking function using floor sensors; the details of the human tracking function are described in Section 5.4.4.3.

5.4.4.3 Floor Sensor

Here we detected people's positions with floor sensors whose resolution is 10×10 cm². The size of one floor sensor is 50×50 cm² and contains 25 sensor units. Since the floor sensors are occlusion free and robust to changes in lighting conditions, they estimate people's positions better than such sensors as ceiling cameras or laser range finders. The floor sensor output is 1 or 0; it is either detecting pressure or it is not. Floor sensors are connected with each other through an RS-232C interface and have a sampling frequency of 5 Hz.

We installed a human tracking function that can estimate the positions of multiple people with floor sensors [10]. This function assigns temporary IDs to detected people such as "TA" and "TB" that are used for making hypotheses. Figure 5.30 shows interaction between a robot and people (left) and output of the floor sensors with IDs from the RFID tag reader (right). The robot can estimate the reactive points itself with its position information.

5.4.5 Hypotheses-Based Mechanism

In this section, we describe how to make and verify hypotheses through interaction between a robot and people. Figure 5.31 shows the interaction flow of the robot. To the first person who gives his or her RFID tag to the

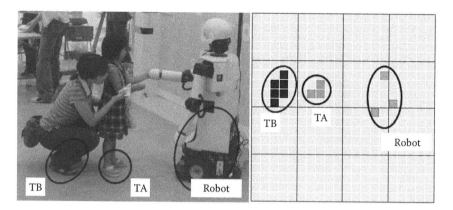

FIGURE 5.30
Floor sensors.

robot (Figure 5.31, step 1), there are four main kinds of robot behavior: greeting (Figure 5.31, step 2), self-introduction (Figure 5.31, step 3), shaking hands (Figure 5.31, step 4), and asking about favorite foods (Figure 5.31, step 5). We call the first and second RFID tags "A-tag" and "B-tag," respectively.

When the robot detects a tag, it formulates a hypothesis using sensor outputs to identify the visitors (Figure 5.31, step 1). The hypothesis reflects the person who first handed his or her RFID tag to the robot. If the hypothesis reliability is high, the robot behaves based on this hypothesis. Thus, the robot greets the person and makes eye contact. If the reliabilities of the hypotheses are less than the defined threshold, the robot greets the person and bows without changing its direction (Figure 5.31, step 2). Further, the robot will greet the second person who gives his or her RFID tag to the robot before asking the first person about the behavior.

To make hypotheses, we integrated the sensor information using a Bayesian network [11]. Such an approach allows us to use people's behavior patterns as parameters to estimate hypothesis accuracy. This advantage is crucial to make hypotheses based on factual interaction data.

The robot checks the hypotheses by interacting with the visitors and estimates the change in the relationship distance between itself and the two people after greetings (Figure 5.31, step 2) and shaking hands (Figure 5.31, step 4). The robot also confirms the existence of the first person's face by height and distance information with the self-introduction behavior (Figure 5.31, step 3). Thus, the robot gradually improves the accuracy of the hypotheses by interaction with these two people.

5.4.5.1 Making a Hypotheses Mechanism

In this study, we designed three kinds of Bayesian networks to identify people and to verify the reliability of the selected hypotheses.

Step 1 Robot detects a tag

Step 2 Robot greets people

Step 3 Robot finds faces of people

Step 4 Robot shakes hand

Step 5 Robot asks about favorite foods

FIGURE 5.31
Interaction flow between people and robot.

5.4.5.1.1 Bayesian Network for Identifying Interacting People Using Multi-sensor Information

Figure 5.32 shows the Bayesian network developed to identify interacting people using multiple sensor information. This Bayesian network is used to make hypotheses when the robot detects an RFID tag the first time in an interaction with two people (Figure 5.31, step 1). The following are the details of each variable of each node.

5.4.5.1.1.1 Tag This variable has two values: *child-tag* and *adult-tag*. If the registered age information of the tag detected by the robot is less than 16,

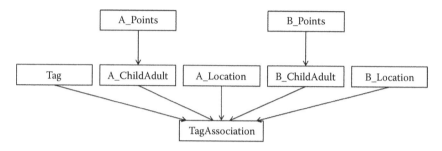

FIGURE 5.32
Bayesian network for identifying interacting people.

this variable result is "child-tag=1" and "adult-tag=0." If not, the result is "child-tag=0" and "adult-tag=1." Thus, this variable depends on the detected tag information.

(a) Step 1 Robot detects a tag (b) Step 2 Robot greets people (c) Step 3 Robot finds faces of people

(d) Step 4 Robot shakes hands (e) Step 5 Robot asks about favorite foods

5.4.5.1.1.2 A_Points and B_Points   These variables have four values, *smallest, small, large,* and *largest,* that change based on the number of reactive points of the floor sensors from all people. We defined the relationships between the variables and the number of reactive points, as shown in Equation 1 (*n* means the number of reactive points). A_Points (or B_Points) mean the number of reactive points of a person with temporary ID TA (or TB). Thus, these variables depend on the number of reactive points of the floor sensors

$$\begin{matrix} A\_Points \\ \\ B\_Points \end{matrix} = \begin{cases} "smallest" = 1, & other = 0 & 0 \le n \le 2 \\ "small" = 1, & other = 0 & 3 \le n \le 5 \\ "large" = 1, & other = 0 & 6 \le n \le 8 \\ "largest" = 1, & other = 0 & 9 \le n \end{cases} \quad (5.14)$$

5.4.5.1.1.3 A_ChildAdult and B_ChildAdult   These variables have two values: *child* and *adult,* which change based on learning datasets. The relationship between the values of A_Points (or B_Points) and the age information of person-A (or person-B) are calculated with gathered interaction data between the robot and the people in the shopping center. If the age of a person is less than 16, the person is regarded as a *child.*

5.4.5.1.1.4 A_Location and B_Location   These variables have three values: *near, middle,* and *far,* which change based on the distance between the robot and a person estimated by the floor sensors. We defined the relationships between the variables and the distance relationship, as shown in Equation 2. A_Location (or B_Location) means the distance between the robot and a person with temporary ID TA (or TB). Thus, these variables depend on the distance relationships between the robot and each person.

$$\begin{array}{c} A\_Location \\ = \\ B\_Location \end{array} \begin{cases} "near" = 1, \ other = 0 & r \leq 0.7 \ and \ d \leq 30 \\ "middle" = 1, \ other = 0 & r \leq 1.0 \ and \ "near"! = 1 \\ "far" = 1, \ other = 0 & r > 1.0 \end{cases} \quad (5.15)$$

Here, r represents the distance between the robot and the person [m], and d represents the direction to the person with respect to the coordinates fixed to the robot.

5.4.5.1.1.5 TagAssociation This variable has two values: A and B, which change based on learning datasets. The relationship between the values of the parent nodes and a factual tag holder is calculated with gathered interaction data between the robot and people in the shopping center.

Calculated value A (or B) of "TagAssociation" is stored as *PA1* (or *PB1*) and is used for verifying the hypothesis process. *PA1* represents the probability that a person with temporary ID TA has an A-tag at step 1 of Figure 5.31. *PB1* represents the probability that a person with temporary ID TB has an A-tag at step 1 of Figure 5.31.

5.4.5.1.2 Bayesian Network for Verifying Hypotheses Based on Distance Relationship

Figure 5.33 shows a developed Bayesian network for verifying hypotheses using distance relationships estimated by floor sensors. It is used to verify the reliability of the hypotheses after greeting or handshaking behaviors (Figure 5.31 and steps 2 and 4). The following are the details of each variable of each node.

5.4.5.1.2.1 A_Location and B_Location These nodes, which represent the distance between the robot and a person, have three states: "near," "middle,"

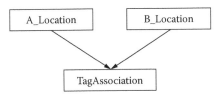

FIGURE 5.33
Bayesian network for verifying hypotheses based on distance relationship.

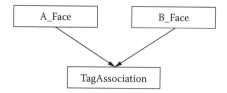

FIGURE 5.34

Bayesian network for verifying hypotheses based on face existence.

and "far." The meanings of these variables are identical to their meanings for the Bayesian network used to identify interacting people.

5.4.5.1.2.2 TagAssociation   These variables have two values: A and B, which change based on the values of the parent nodes. We expected that a person will approach a robot when it greets or shakes hands with the person being addressed. These variables change based on the values of the parent nodes; we designed a Bayesian network that only uses A_Location and B_Location. The meanings of these variables are identical to their meanings for the Bayesian network used to identify interacting people.

The calculated value A (or B) of "TagAssociation" is stored as *PA2(4)* (or *PB2(4)*) and is used for verifying the hypothesis process. *PA2(4)* represents the probability that a person with temporary ID TA has an A-tag at step 2(4) of Figure 5.31. *PB2(4)* represents the probability that a person with temporary ID TB has an A-tag at step 2(4) of Figure 5.31.

5.4.5.1.3 Bayesian Network for Verifying Hypotheses Based on Distance Relationship

Figure 5.34 shows the Bayesian network developed for verifying hypotheses based on face existence estimation by the robot's face detection function. This Bayesian network is used to verify the reliability of the hypotheses after the face-searching behavior (Figure 5.31, step 3). The following are the details of each variable of each node.

5.4.5.1.3.1 A_Face and B_Face   These variables have two values: *face* and *none*, which change based on the results of the face detection function. If the robot detects a face when looking at a person with temporary ID TA (or TB), the value of A_Face (or B_Face) is "face=1" and "none=0."

5.4.5.1.3.2 TagAssociation   These variables have two values: *A* and *B*, which change based on the values of the parent nodes: A_Face and B_Face. If the robot only detected the face of one side of a person, the reliability of hypotheses will be updated. The meanings of these variables are identical to their meanings for the Bayesian network used to identify interacting people.

The calculated value A (or B) of TagAssociation is stored as *PA3* (or *PB3*) and is used for verifying the hypotheses process. *PA3* represents the probability that a person with temporary ID TA has an A-tag at step 3 of Figure 5.31. *PB3* represents the probability that a person with temporary ID TB has an A-tag at step 3 of Figure 5.31.

5.4.5.2 Verifying Hypotheses Mechanism

For verifying hypotheses, we designed three kinds of behavior: greeting, self-introduction, and shaking hands. These verification behaviors are controlled with the reliabilities of the hypotheses. For example, when the robot has positively identified person A, it prefers such conclusive behavior as a greeting and eye contact. On the other hand, when the robot has doubts about person A, it prefers such nonconclusive behavior as a greeting and a bow without eye contact. By behaving nonconclusively and sensing a change in the person's behavior, the robot tries to solve ambiguity.

During step 1 of Figure 5.31, the robot makes two hypotheses: *PA* and *PB*. *PA* represents the probability that the person with temporary ID TA has an A-tag, and *PB* represents the probability that the person with temporary ID TB has an A-tag. At this step, these values are calculated using *PA1* and *PB1* (Equation 5.16):

$$PA = PA1$$
$$PB = PB1 \tag{5.16}$$

During steps 2 to 4 of Figure 5.31, the robot makes hypotheses to verify the hypotheses of PA and PB by interacting with PA and PB. The hypotheses of PA and PB are updated and normalized at each steps (Equation 5.17):

$$PA = PA * PA_i / (PA * PA_i + PB * PB_i)$$
$$PB = PB * PB_i / (PA * PA_i + PB * PB_i) \qquad i = number\ of\ step \tag{5.17}$$

When the hypothesis reliability is increased by verifying behavior, the robot behaves conclusively. If it behaves conclusively with people owing to an incorrect hypothesis, it apologizes to the first person and continues the interaction. Keeping multiple hypotheses by verifying them enables the robot to interact and correct interaction failures.

5.4.6 Performance Evaluation

5.4.6.1 Settings

Environment We evaluated the proposed method with a test field at the AEON Takanohara shopping center from August 30 to 31, 2007. Visitors to the shopping center were mainly families who were shopping. Figure 5.35

FIGURE 5.35
Robot and floor sensors in shopping center.

shows the environment; the robot and the floor sensors were installed at the main passage of the shopping center that is visited by more than 30,000 people every day.

Participants If a visitor decides to register as part of our project, we gather such personal data as name, height, and incorporate such information and an RFID tag's ID. We gave one tag to 14 pairs of visitors and two tags to 10 pairs. In this environment, we also gathered 27 pieces of interaction data between the robot and the pairs for the learning of the Bayesian network. These pairs were mainly families and friends. When a pair consisted of a parent and a child, we asked the parent not to bend because we anticipated that the parent would bend to adjust to the child's height. Also, we asked participants to hand the tag to the robot after both participants stepped on the floor sensors.

5.4.6.2 Result

Figure 5.36 shows the success rate of identification when the robot interacted with participants. In this method, the identification accuracy is 79.2% (a pair having one tag is 92.9%; a pair having two tags is 60%). The results indicate that the proposed method accurately identified when only one person had a tag. However, the proposed method sometimes failed to identify two persons when they had two tags.

To investigate the accuracy of the verifying hypothesis mechanism, we compared the performance using the proposed method without the

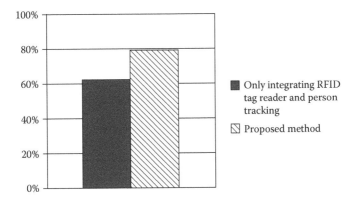

FIGURE 5.36
Success rate of identification. (a) The hypothesis' reliability that TB has the tag is high. (b) The hypotheses' reliabilities are changing. (c) The hypothesis' reliability that TA has the tag is high.

mechanism. Figure 5.36 also shows the success rate of identification without the verifying hypothesis mechanism in which identification accuracy is 62.5% (a pair having one tag is 71.4%; a pair having two tags is 50%), indicating that the hypothesis-based system corrected interaction failures and solved the ambiguous situations through interaction.

Figure 5.37 shows a case where the robot corrected a failure through interaction. The robot shakes hands while making eye contact based on the hypothesis, but the hypothesis is wrong (Figure 5.37a). At the same time, the two persons changed positions; the robot detected the change of the distance relationship (Figure 5.37b). Then the reliabilities of the hypotheses were changed, and the robot behaved based on the changed hypotheses; the robot apologized and continued to interact with them.

5.4.7 Discussion

5.4.7.1 HRI Contributions

Based on the proposed method, the developed system identified interacting people through verifying behavior with 79.2% accuracy. In addition, we compared the accuracy of the identification between the proposed method and a method that only integrated floor sensors and passive-type RFID tag readers. As a result, the proposed method improved the successful rate by 16.7%, indicating that the verifying hypothesis mechanism of the proposed method enables the robot to resolve ambiguous situations when identifying multiple people in a real environment.

We believe the proposed method can be applied to resolve other ambiguous situations in HRI such as complicated guiding or explaining multi-product sets with similar names. In such situations, a robot with the proposed

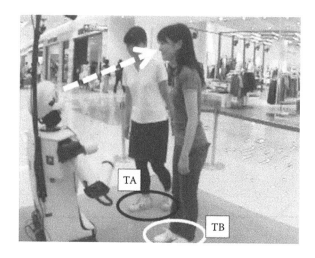

(a)

(b)

FIGURE 5.37
An example of verification behavior.

method can resolve ambiguous situations through interaction. Furthermore, it can behave more intelligently by gathering interaction data for the long term. These interaction data improve the accuracy of hypotheses for making and verifying the hypothesis process.

In addition, a person identification function was developed with the technology of a contact identification device that indicates that the system considers the privacy issues of person identification. This advantage is important for future human robot interaction.

(c)

FIGURE 5.37
An example of verification behavior.

5.4.7.2 Performance Improvement Approach

The robot with the proposed method identifies interacting people with 92.9% accuracy when the interacting people only have one tag. However, when they have two tags, the identification accuracy was 60%.

We assume that the main reason for failure is the differences in the interacting people's behavior when they have one or two tags. When the people have two tags, they often changed their positions to give their tag to the robot. When the robot addresses a person, the other person tried to give his or her tag to it, ignoring the robot's behavior. We believe that the hypothesis-based system could not correct the interaction failures due to such behavior on the part of the interacting people. To solve these problems, more learning datasets are needed that must include the particular situations in which the robot interacted with two people with tags.

5.4.8 Conclusion

We proposed a method with which a communication robot simultaneously interacts with two persons by identifying them with a passive-type RFID reader and floor sensors. The method allows the robot to consider multiple hypotheses about the interpretation of sensor information and gradually improves the accuracy of the hypotheses through interaction with people. A Bayesian network, which was used to integrate the output of these sensors, enables us to use people's behavior patterns as parameters to estimate the accuracy of the hypotheses. We installed the proposed method in a communication robot and verified its efficiency in a shopping center. Experimental results revealed that the system successfully identified 79.2% of the visitors. In particular, the proposed method for behavior control improved the success rate by 16.7%.

We believe that the proposed method contributes to the achievements of a communicative robot that simultaneously identifies and interacts with many people in a real environment because it allows the robots to interact with people based on interaction data gathered in each environment. In addition, the hypothesis-based system can be applied not only to identification problems but also to such interaction problems as speech recognition, sensor processing, and so on.

Acknowledgments

We wish to thank the staff of the AEON Co. Ltd. for their kind cooperation. We also wish to thank the following ATR members for their helpful suggestions and cooperation: Daisuke Sakamoto, Dylan F. Glass, Zenta Miyashita, Yoshii Akira, Atsushi Izawa,

References

1. Kanda, T., Hirano, T., Eaton, D., and Ishiguro, H. Interactive robots as social partners and peer tutors for children: A field trial, *Human Computer Interaction*, Vol. 19, No. 1–2, pp. 61–84, 2004.
2. Shiomi, M., Kanda, T., Ishiguro, H., and Hagita, N. Interactive humanoid robots for a science museum, *1st Annual Conference on Human-Robot Interaction*, pp. 305–312, 2006.
3. Nourbakhsh, I., Bobenage, J., Grange, S., Lutz, R., Meyer, R., and Soto, A. An affective mobile educator with a full-time job, *Artificial Intelligence*, 114 (1–2), pp. 95–124, 1999.
4. Siegwart, R., et al. Robox at Expo. 02: A large scale installation of personal robots, *Robotics and Autonomous Systems*, 42, pp. 203–222, 200, 2003.
5. Hayashi, K., Sakamoto, D., Kanda, T., Shiomi, M., Koizumi, S., Ishiguro, H., Ogasawara, T., and Hagita, N. Humanoid robots as a passive-social medium—a field experiment at a train station, *ACM 2nd Annual Conference on Human-Robot Interaction (HRI2007)*, pp. 137–144, 2007.
6. Want, R., Hopper, A., Falcao, V., and Gibbons, J. The active badge location system, *ACM Transactions on Information Systems*, pp. 91–102, 1992.
7. Lao, S., and Kawade. M. Vision-based face understanding technologies and their applications, *SINOBIOMETRICS*, pp. 339–348, 2004.

8. Vogt, H. Efficient object identification with passive RFID tags, In *Proceedings of International Conference on Pervasive Computing (Pervasive 2002)*, Aug. 2002.

9. Sim, T., Zhang, S., Janakiraman, R., and Kumar, S. Continuous verification using multimodal biometrics, *IEEE Transactions on Pattern Analysis and Machine Intelligence*, April 2007 (to appear).

10. Murakita, T., Ikeda, T., and Ishiguro, H. Human tracking using floor sensors based on the Markov Chain Monte Carlo Method, *International Conference on Pattern Recognition*, pp. 917–920, 2004.

11. Jensen, F. V. *Bayesian Networks and Decision Graphs*, Springer, 2001.

12. Fransen, B. R., Morariu, V. I., Martinson, E., Blisard, S., Marge, M., Thomas, S., Schultz, A., and Perzanowski, D. Using vision, acoustics, and natural language for disambiguation, *ACM 2nd Annual Conference on Human-Robot Interaction (HRI2007)*, pp. 73–80, 2007.

13. Inamura, T., Inaba, M., and Inoue, H. A dialogue control model based on ambiguity evaluation of users' instructions and stochastic representation of experiences, *Journal of Robotics and Mechatronics*, Vol. 17, No. 6, pp. 697–704, 2005.

14. Tajika, T., Miyashita, T., Ishiguro, H., and Hagita, N. Automatic categorization of haptic interactions—what are the typical haptic interactions between a human and a robot?, *IEEE International Conference on Humanoid Robots (Humanoids)*, Dec. 2006.

15. Ishiguro, H., Ono, T., Imai, M., Maeda, T., Kanda, T., and Nakatsu, R. Robovie: An interactive humanoid robot, *Int. J. Industrial Robot*, Vol. 28, No. 6, pp. 498–503, 2001.

16. Shiomi, M., et al. Face-to-face interactive humanoid robot, *IEEE/RSJ International Conference on Intelligent Robots and Systems*, pp. 1340–1346, Sep. 2004.

5.5 Friendship Estimation Model for Social Robots to Understand Human Relationships

Takayuki Kanda and Hiroshi Ishiguro

ATR Intelligent Robotics Laboratories

ABSTRACT

This chapter reports the friendship estimation model we designed for social robots that understand human social relationships. Our interactive robot autonomously interacts with humans with its human-like body properties, and as a result, induces the humans' friendly group behavior upon direct interaction. Based on these features, as well as inspired by a survey in psychology research on friendship, we propose a friendship estimation model for an interactive robot. The capability provided by such a model is probably essential for interactive robots to establish social relationships with humans.

The results of an experiment demonstrate that the fundamental part of the estimation model functions effectively.

5.5.1 Introduction*

Recent progress in robotics has brought with it a new research direction known as "interaction-oriented robots." These robots are different from traditional task-oriented robots, such as industrial robots, that perform certain tasks in limited applications. Interaction-oriented robots are designed to communicate with humans and to participate in human society. We are trying to develop such an interaction-oriented robot that will behave as a partner in people's daily lives. We believe these robots will not only be used for entertainment but also provide communication support such as route guidance and mental support tasks.

Several researchers, such as Aibo and Kismet [1], are endeavoring to realize interaction-oriented robots. Moreover, several research works are exploring the possible applications of interactive robots. Shibata et al. successfully employed a seal-like pet robot, Paro, for the mental care of elderly persons [2]. Dautenhahn et al. has used a simple interactive robot for autism therapy [3]. These research efforts seem to be devoted to social robots that are embedded in human society.

The research question we are struggling to solve is, "How can an interaction-oriented robot participate in human daily life, establish social relationships with humans, and contribute to society?" In other words, our purpose is to realize a peer/partner robot that socially communicates with humans to support their daily lives.

We believe that the social ability of robots will be greatly improved by involving them in human society. The initial tasks of the robots will be limited and perhaps trivial, since the interaction abilities of the current robots are not as high as human infants' and their social skills are very low. However, we can improve the social skills of robots in society by finding various problems that robots will address, by imitating similar development steps as those of human infants.

Currently, robots work in our daily lives as interactive robots, and they are gradually improving their interactive abilities; however, they have not advanced to the stage of social work that requires social communication with more than one person. While previous research works have developed robots' interactive abilities for only one person in front of the robots, we believe it is imperative to improve robots' social ability to make them work in our daily lives, which is the approach indicated by the broken arrow

---

* This chapter is a modified version of a previously published paper Takayuki Kanda and Hiroshi Ishiguro, *Friendship estimation model for social robots to understand human relationships*, IEEE International Workshop on Robot and Human Communication (ROMAN2004), pp. 539–544, 2004, edited to be comprehensive and fit with the context of this book.

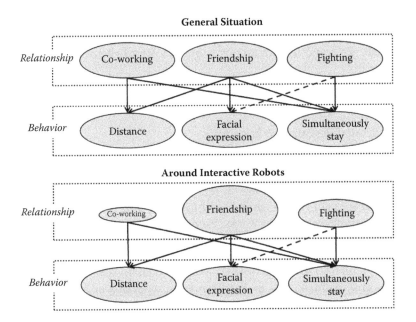

FIGURE 5.38
Relations between social relationships and group behavior.

in Figure 5.38. We believe that robots' tasks will emerge according to the improvement in robots' abilities, even if current robots are equipped with little skill to accomplish useful tasks in human society.

We are pursuing this approach of making robots gradually work in our daily lives to improve their social abilities as well as to explore the possible tasks of the robots. The first step of the approach was a field trial in an elementary school where interactive robots behaved as peer tutors of a foreign language, English, as reported in [4]. The Robovie robot was equipped with a person-identification function to distinguish children, such as for calling the names of children, and simultaneously interacted with more than one child. As a result, it demonstrated the possibility that interactive robots can motivate children to learn a foreign language through interaction with robots. Meanwhile, we have observed group behavior among friends around the robot. For instance, a boy and his friend counted how many times the robot called their respective names, and the boy whose name was called more often proudly told his friend that the robot preferred him. If the robot could understand their friendship, it could promote interaction with the boys and interaction between the boys. That is, the ability of friendship estimation would enable robots to mediate the interaction between humans. Moreover, friendship is tightly connected to social relationships (described in the next section in detail). Thus, this friendship estimation is essential for accomplishing more general social relationship estimation, which might provide a future social robot that could help to solve the bullying problem or the problem of isolated children. In this chapter, we report our approach to estimating human friendship by using an interactive robot, an ability that is probably essential for interactive robots to establish social relationships with humans.

5.5.2 Friendship Estimation Model from Observation

5.5.2.1 Related Research on Friendship

It is a well-grounded finding from psychological research that children at a very young age engage in dyadic relationships, for example, in the form of pretend play, that then increase in size and complexity with age, forming many different peer relationships in the form of social networks. As children gradually establish social networks, each child gets a different social status, such as popular, average, isolated, and rejected [5,6].

A sociometric test has been used to investigate peer relationships and social networks, which let a human directly answer the name of others whom he or she likes and dislikes. This method has been validated as a reliable assessment of human peer relationships. It categorizes each child's social status into one of such groups: popular, average, neglected, and rejected [7,8]. It has been widely used to determine the relationships in a classroom or a company.

On the other hand, observation-based methods have been developed for understanding peer relations and social status. Children form a group and behave with the group, along with their friendly relationships. Children usually play with peers, while boys tend to play in groups and girls tend to play with only one other girl [9]. Ladd et al. investigated the relationships between observed group behavior and their relationships. They coded videotape of children's play with four behavioral measures: cooperative play, rough play, unoccupied, and teacher orientation. This revealed that cooperative play was associated with positive nominations, while rough play was related to negative nominations. In addition, they revealed that past behavior successfully predicted the current peer status; for example, time spent in cooperative play was a significant predictor of positive nomination [6]. Coie et al. investigated the difference between popular and rejected children in terms of their behavior and revealed the relationships between rejected children and their aversive behaviors [10]. We believe these findings support the possibility that social robots can recognize humans' peer relationships and social status by observing their group behavior.

5.5.2.2 Related Research on Sensor-Based Observation of Humans' Interactions

Our research approach is to recognize human social relationships by observing group behavior in the presence of an interactive robot, which is closely related to the research works that attempt to analyze human interaction from sensors embedded in environments or attached to humans. Recently, several research works have attempted to automatically observe and analyze large-scale human behaviors by using virtual reality, wearable computing, and ubiquitous sensing technologies.

Velde et al. proposed a mixed-reality approach for supporting human activity at conferences [11], which is known as the COMRIS project. Here, a parrot-like physical agent on human shoulders supplies information connected to a backbone information infrastructure. At the same time, this system measures interactions in large groups occurring spontaneously. With recent ubiquitous sensing technology, Sumi et al. developed an ubiquitous sensor environment for capturing and analyzing human physical group interaction by using infrared sensors that are good at sensing directions such as eye gaze [12]. By using the sensing system, Bono et al. analyzed human social behaviors during a poster session in an exhibition, which revealed the role of nonverbal cues for conversation between an exhibiter and visitors as well as the possibility of detecting their preferences for the exhibited posters [13].

On the other hand, several research works have attempted to analyze human behaviors toward interactive robots. For example, Dautenhahn et al. analyzed children's eye gaze and contact time for an interactive robot in the AURORA project [3], which aims to apply interactive robots to helping autistic children acquire social skills. Moreover, a motion capturing system enables us to automatically capture human embodied behavior during human–robot interaction. Kanda et al. measured the spatial and temporal synchrony between humans and an interactive robot to predict humans' subjective impressions of the robot [14].

5.5.2.3 Friendship Estimation Model

Human behavior is largely based on social relationships, which can be in the form of dyadic relationships, known as friendship, or larger groups known as social networks, where there are complex peer relationships among different individuals. Since the previous research works have proved the correlations between children's group behavior and their relationships [6,9,10], we believe we can estimate their peer relationships and social networks from observation of their group behavior. We focused on the *estimation* of peer relationships, which are the fundamental parts of the social network, as an early attempt at *recognition* of peer relationships and social network. Yet it is not through *recognition* (finding all correct information accurately) but through *estimation* (partially finding correct information with moderate accuracy) that robots can utilize obtained information to further promote human–robot interaction.

The basic idea is that "a robot autonomously interacts with several children simultaneously to cause their spontaneous group behavior, and the group behavior is observed to recognize their relationships," which is also our hypothesis to verify. Our friendship estimation model is based on the association of social group behavior and social relations, which is inspired by previous psychological research such as the works mentioned above. In general, humans' social relationships affect group behavior, such as accompanying distance among members, facial expressions during conversation, and so forth. For instance, a human is accompanied by another friendly human but does not willingly approach a disliked human (accompanying and close distance). Sometimes, such unfriendly relations cause a quarrel or fight (distance will be close, but facial expression will be far from friendly). Meanwhile, official relationships rather than private ones sometimes cause nonspontaneous group behavior. For instance, a teacher may organize co-working activity such as "children collaborate to carry a heavy box." Figure 5.38 (left) describes these examples of the associations between group behaviors and peer relations in general situations.

On the other hand, according to our hypothesis, an interactive robot mostly causes spontaneous friendly behaviors. In fact, we observed such situations, where a child is accompanied by his or her friend when interacting with the robot as shown in Figure 5.39. We are going to verify this hypothesis in this chapter later. Thus, we believe we can estimate such friendly relationships by simply observing group behavior. This idea is described in Figure 5.40 (right). As the beginning step for the estimation, we only utilize "accompanying" behavior that can be recognized by using a wireless tag system.

5.5.2.4 Algorithm

Figure 5.40 (left) indicates the mechanism of the friendship estimation. From a sensor (in this case, wireless ID tags and receiver), the robot constantly obtains the IDs (identifiers) of individuals who are around it. The robot continuously accumulates its interacting time with person A (T_A) and the time that person A and B simultaneously interact with it (T_{AB}, which is equivalent to T_{BA}). We define the estimated friendship from person A to B (*Friend(A→B)*) as

$$\text{Friend}(A{\rightarrow}B) = \text{if } (T_{AB}/T_A > T_{TH}), \tag{5.18}$$

$$T_A = \Sigma \text{ if (observe(A) and } (St < S_{TH})) \cdot \Delta t, \tag{5.19}$$

$$T_{AB} = \Sigma \text{ if (observe(A) and observe(B) and } (St < S_{TH})) \cdot \Delta t, \tag{5.20}$$

where *observe(A)* becomes true only when the robot observes the ID of person A, *if()* becomes 1 when the logical equation inside the parentheses is true (otherwise, 0), and T_{TH} is a threshold of simultaneous interaction time. We also prepared a threshold S_{TH}, and the robot only accumulates T_A and T_{AB} so

FIGURE 5.39
Scenes of a child accompanied by a friend interacting with a humanoid robot.

that the number of persons simultaneously interacting at time t (St) is less than S_{TH} (Equations 5.19 and 5.20). In our trial, we set Δt to one second.

5.5.3 Robovie: An Interactive Humanoid Robot

5.5.3.1 Hardware of Interactive Humanoid Robot

Figure 5.41 shows the humanoid robot "Robovie" [15]. The robot is capable of human-like expression and recognizes individuals by using various actuators and sensors. Its body possesses highly articulated arms, eyes, and a head, which were designed to produce sufficient gestures to communicate

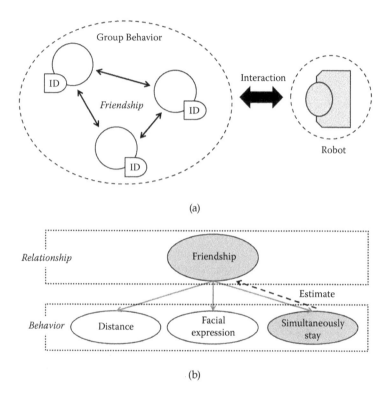

(a)

(b)

FIGURE 5.40
Current estimation model for friendship.

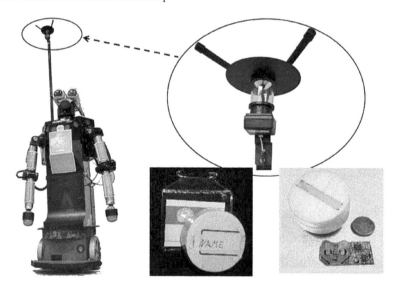

FIGURE 5.41
Robovie (left) and wireless tags.

effectively with humans. The sensory equipment includes auditory, tactile, ultrasonic, and vision sensors, which allows the robot to behave autonomously and to interact with humans. All processing and control systems, such as the computer and motor control hardware, are located inside the robot's body.

5.5.3.2 *Person Identification with Wireless ID Tags*

To identify individuals, we used a wireless tag system capable of multi-person identification by partner robots (detailed specification and system configuration is described in [16]). Recent RFID (radio frequency identification) technologies have enabled us to use contactless identification cards in practical situations. In this study, children were given easy-to-wear nameplates (5 cm in diameter), in which a wireless tag was embedded. A tag (Figure 5.41, lower-right) was periodically transmitted its ID to the reader installed on the robot. In turn, the reader relayed received IDs to the robot's software system. It was possible to adjust the reception range of the receiver's tag in real time by software. The wireless tag system provided the robots with a robust means of identifying many children simultaneously. Consequently, the robots could show some human-like adaptation by recalling the interaction history of a given person.

5.5.3.3 *Interactive Behaviors*

"Robovie" features a software mechanism for performing consistent interactive behaviors (detailed mechanism is described in [17]). The objective behind the design of Robovie is that it should communicate at a young child's understanding level. One hundred interactive behaviors have been developed. Seventy of them are interactive behaviors such as shaking hands, hugging, playing paper-scissors-rock, exercising, greeting, kissing, singing, briefly conversing, and pointing to an object in the surroundings. Twenty are idle behaviors such as scratching the head or folding the arms, and the remaining 10 are moving-around behaviors. In total, the robot could utter more than 300 sentences and recognize about 50 words.

Several interactive behaviors depended on the person identification function. For example, there was an interactive behavior in which the robot called a child's name if that child was at a certain distance. This behavior was useful for encouraging the child to come and interact with the robot. Another interactive behavior was a body-part game, where the robot asked a child to touch a body part by saying the part's name.

The interactive behaviors appeared in the following manner based on some simple rules. The robot sometimes triggered the interaction with a child by saying, "Let's play, touch me," and it exhibited idling or moving-around behaviors until the child responded; once the child reacted, it continued performing friendly behaviors for as long as the child responded. When the child stopped reacting, the robot stopped the friendly behaviors, said "good bye," and restarted its idling or moving-around behaviors.

5.5.4 Experiment and Result

We conducted a field experiment in an elementary school for two weeks with the developed interactive humanoid robot, which was originally designed to promote children's English learning. As we reported in [4], the robots had a positive effect on the children. In this paper, we use the interaction data during that trial as a test-set of our approach to reading friendship from the children's interaction.

5.5.4.1 Method

We performed an experiment at an elementary school in Japan for two weeks. Subjects were sixth-grade students from three different classes, totaling 109 students (11–12 years old, 53 male and 56 female). There were nine school days included in those two weeks. Two identical robots were placed in a corridor that connects the three classrooms (Figure 5.42). Children could freely interact with both robots during recesses (in total, about an hour per day), and each child had a nameplate with an embedded wireless tag so that each robot could identify the child during interaction.

We administered a questionnaire that asked the children to write down the names of their friends. This obtained friendship information was collected for comparison with the friendship relationships estimated by our proposed method.

5.5.4.2 Results for Frequency of Friend-Accompanying Behavior

As we compared the questionnaire on friendships and the interacting time with the robot, we found a higher frequency with which children interacted with the robot in the company of his or her friend (see Figures 5.43). Seventy-two percent of their interaction time with the robot was in the

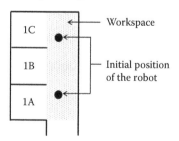

FIGURE 5.42
Environment of elementary school.

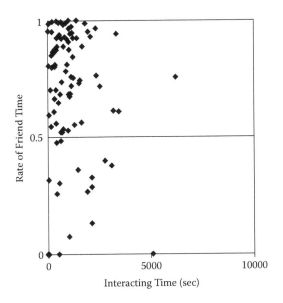

FIGURE 5.43
Frequency of friend-accompanying behavior.

company of one or more friends. We believe that this result supports our hypothesis that "our interactive robot mostly causes friendly accompanying behavior of children around it rather than the other behaviors associated with non-friendly relationships, such as hostile, dislike, co-working." This implies that we can estimate their friendship by even simply observing their accompanying behavior.

5.5.4.3 Results for Friendship Estimation

Since the number of friendships among children was fairly small, we focused on the appropriateness (coverage and reliability) of the estimated relationships. This is similar to the evaluation of an information retrieval technique such as a Web search. Questionnaire responses indicated 1,092 friendships among a total of 11,772 relationships; thus, if we suppose that the classifier always classifies a relationship as a non-friendship, it would obtain 90.7% correct answers, which means the evaluation is completely useless. Thus, we evaluate our estimation of friendship based on reliability and coverage, which are defined as follows:

Reliability = number of correct friendships in estimated friendships/number of estimated friendships

Coverage = number of correct friendships in estimated friendship/number of friendships from the questionnaire

TABLE 5.4

Estimation Results with Various Parameters

Coverage Reliability		T_{TH} (Simultaneously Interacting Time)					
		0.3	0.2	0.1	0.05	0.01	0.001
S_{TH} (num. of simultaneously Interacting children)	2	0.01	0.02	0.03	0.04	0.04	0.04
		1.00	0.93	0.79	0.59	0.54	0.54
	5	0.00	0.02	0.06	0.11	0.18	0.18
		1.00	1.00	0.74	0.47	0.29	0.28
	10	0.00	0.00	0.04	0.13	0.29	0.31
		—	1.00	0.74	0.46	0.23	0.20

('—' indicates that no relationships were estimated, so reliability was not calculated)

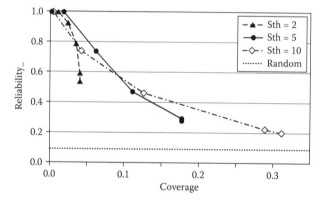

FIGURE 5.44
Illustrated estimation results.

Table 5.4 and Figure 5.44 indicate the results of estimation with various parameters (S_{TH} and T_{TH}). In Figure 5.44, random represents the reliability of random estimation where we assume that all relationships are friendships (since there are 1,092 correct friendships among 11,772 relationships, the estimation obtains 9.3% reliability at any coverage). In other words, random indicates the lower boundary of estimation. Each of the other sketch lines in the figure represents the estimation result with different S_{TH}, which has several points corresponding to different T_{TH}. There is obviously a trade-off between reliability and coverage, which is controlled by T_{TH}; S_{TH} has a small effect on the trade-off, S=5 mostly performs better estimation of the friendship, and S=10 performs better estimation when coverage is more than 0.15. As a result, our method successfully estimated 5% of the friendship relationships with greater than 80% accuracy (at "S_{TH}=5") and 15% of them with nearly 50% accuracy (at "S_{TH}=10") (these early findings on friendship estimation, which are reported in this subsection, have already appeared in our previous paper [18]).

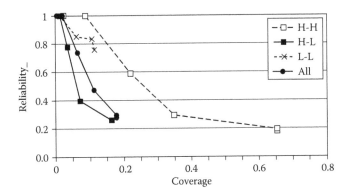

FIGURE 5.45
Relationships between friendship estimation and frequency of interaction with robot (at Sth = 5).

5.5.4.4 Results for Comparison between Frequency of Interaction with the Robot and Estimation Result

We analyzed the results in detail to determine how we could improve the estimation. Figure 5.45 shows a comparison between frequency of interaction with the robot and estimation results. We classified the children into two groups as higher-frequency half of 109 children (referred to H) and lower-frequency half (L) according to their interacting time with the robot. We then illustrated the estimation performance among H-H (friendship between children within the H group), H-L (friendship between a child in the H group and a child in the L group), and L-L (friendship between children within the L group). The comparison's results revealed that our method better estimated the relationships in the H-H group. In other words, the estimation is more accurate for the relationships between children who often appeared around the robot. In contrast, the upper boundary of the coverage is more limited for the relationships between children who did not often appear around it. (As Figure 5.43 shows, there is no obvious correlation between frequency of interaction with the robot and the rate of friend-accompanying behavior; thus, this does not cause the difference in estimation between the H-H group and others.) We believe these findings also support the effectiveness of our estimation model, since it seems that the estimation will become more accurate with an increase in the amount of observed data on inter-human relationships around the robot.

5.5.5 Conclusions

We proposed a friend estimation model for a social robot that interacts with humans and verified the fundamental component of the model by field experiment. In the field experiment, two identical interactive humanoid robots were placed in an elementary school for two weeks, where children freely interacted with the robots during recesses. These interactive humanoid robots identified individual children by using a wireless tag system, which is utilized for recording individual and friend-related interaction time as well as for promoting interaction by such actions as calling their names. The result suggested that children were mostly accompanied by one of more friend (72% of the total interacting time) and that the robot successfully estimated friendly relationships among children who often appeared around the robot while it showed moderate performance for the others (for example, 5% of all relationships with 80% accuracy). We believe that these early findings will encourage further research into the social skills of social robots as well as the sensing technology for autonomous observation about inter-human and human–robot interaction.

Acknowledgments

We wish to thank the teachers and students at the elementary school for their willing participation and helpful suggestions. We also thank Takayuki Hirano and Daniel Eaton, who helped with this field trial in the elementary school. This research was supported by the NICT of Japan.

References

1. Breazeal, C., and Scassellati, B. (1999). A context-dependent attention system for a social robot, *Proc. Int. Joint Conf. on Artificial Intelligence,* pp. 1146–1151.
2. Shibata, T., and Tanie K. (2001). Physical and affective interaction between human and mental commit robot, *Proc. of IEEE Int. Conf. on Robotics and Automation,* pp. 2572–2577.
3. Dautenhahn, K., and Werry I. (2002). A quantitative technique for analysing robot-human interactions, *IEEE/RSJ Int. Conf. on Intelligent Robots and Systems,* pp. 1132–1138.
4. Kanda, T., Hirano, T., Eaton, D., and Ishiguro, H. (2004). Interactive robots as social partners and peer tutors for children: A field trial, *Journal of Human Computer Interaction,* 19(1&2), 61–84.
5. Gottman, J. M., and Parkhurst, J. (1980). A developmental theory of friendship and acquaintanceship processes. In W. A. Collins (Ed.), *Minnesota Symposia on Child Psychology* (Vol. 13, pp. 197–253), Hillsdale, NJ: Lawrence Erlbaum.
6. Ladd G. W., Price J. M., and Hart C. H. (1990), Preschooler's behavioral orientations and patterns of peer contact: predictive of peer status? In Asher S. R. and J. D. Coie (Eds.), *Peer Rejection in Childhood* (pp. 90–115), Cambridge University Press.

7. McConnell, S. R., and Odom, S. L. (1986). Sociometrics: Peer-referenced measures and the assessment of social competence. In P. S. Strain, M. J. Guralnick, and H. M. Walker (Eds.), *Children's Social Behavior: Development, Assessment, and Modification* (pp. 215–284), Orlando: Academic Press.
8. Asher, S. R., and Hymel, S. (1981). Children's social competence in peer relations: Sociometric and behavioral assessment. In J. D. Wine and M. D. Smye (Eds.), *Social Competence* (pp. 125–157), New York: Guilford.
9. Waldrop, M. F., and Halverson, C. F. (1975). Intensive and extensive peer behavior: Longitudinal and cross-sectional analyses. *Child Development*, Vol. 46, pp. 19–26.
10. Coie, J. D., and Kupersmidt, J. B. (1983). A behavioral analysis of emerging social status in boys' groups. *Child Development*, Vol. 54, pp. 1400–1416.
11. Van de Velde, W. Co-habited mixed realities. *Proc. of IJCAI'97 workshop on Social Interaction and Communityware* (1997).
12. Sumi, Y., Matsuguchi, T., Ito, S., Fels, S., and Mase, K. (2003). Collaborative capturing of interactions by multiple sensors, *Int. Conf. on Ubiquitous Computing (Ubicomp 2003)*, pp. 193–194.
13. Bono, M., Suzuki, N., and Katagiri, Y. (2003). An analysis of non-verbal cues for turn-taking through observation of speaker behaviors, *Joint Int. Conf. on Cognitive Science (CogSci)*.
14. Kanda, T., Ishiguro, H., Imai, M., and Ono, T. (2003). Body movement analysis of human-robot interaction, *Proc. Int. Joint Conf. on Artificial Intelligence*, pp. 177–182.
15. Ishiguro, H., Ono, T., Imai, M., and Kanda, T. (2003). Development of an interactive humanoid robot "Robovie"—an interdisciplinary approach, R. A. Jarvis and A. Zelinsky (Eds.), *Robotics Research*, Springer, pp. 179–191.
16. Kanda, T., Hirano, T., Eaton, D., and Ishiguro, H. (2003). Person Identification and Interaction of Social Robots by Using Wireless Tags, *IEEE/RSJ International Conference on Intelligent Robots and Systems*, pp. 1657–1664.
17. Kanda, T., Ishiguro, H., Imai, M., Ono T., and Mase, K. (2002). A constructive approach for developing interactive humanoid robots, *IEEE/RSJ International Conference on Intelligent Robots and Systems*, pp. 1265–1270.
18. Kanda, T., and Ishiguro, H. (2004). Reading human relationships from their interaction with an interactive humanoid robot, *Int. Conf. on Industrial and Engineering Applications of Artificial Intelligence and Expert Systems (IEA/AIE)*.

5.6 Estimating Group States for Interactive Humanoid Robots

Masahiro Shiomi, Kenta Nohara, Takayuki Kanda,
Hiroshi Ishiguro, and Norihiro Hagita

ABSTRACT

In human–robot interaction, interactive humanoid robots must simultaneously consider interaction with multiple people in real environments such as stations and museums. To interact with a group simultaneously, we must estimate whether a group's state is suitable for the robot's intended task. This paper presents a method that estimates the states of a group of people for interaction between an interactive humanoid robot and multiple people by focusing on the position relationships between clusters of people. In addition, we also focused on the position relationships between clusters of people and the robot. The proposed method extracts the feature vectors from position relationships between the group of people and the robot and then estimates the group states by using Support Vector Machine with extracted feature vectors. We investigate the performance of the proposed method through a field experiment whose results achieved an 80.4% successful estimation rate for a group state. We believe these results will allow us to develop interactive humanoid robots that can interact effectively with groups of people.

5.6.1 Introduction*

The development of humanoid robots has entered a new stage that is focusing on interaction with people in daily environments. The concept of a communication robot continues to rapidly evolve. Soon, robots will act as peers providing psychological, communicative, and physical support. Recently, humanoid robots have begun to operate in such everyday environments as elementary schools and museums [1–5].

We must focus on group dynamics rather than individuals because robots will need to interact with groups of people in daily environments. For example, Figure 5.46 shows a robot interacting with a group in a science museum. When many people gather, the shape of the group reflects its behavior. For example, many people form a line when going through a ticket gate or a circle when conversing with each other [6]. In human–robot interaction, a fan shape spread out from a robot is suitable when the robot is explaining an exhibit to visitors, as shown in Figure 5.46. From this point of view, we believe that a suitable group state exists for each robot task.

However, almost all past works have neglected group states because they have mainly focused on estimating people's positions, behavior, and relationships. Figure 5.47 shows a comparison between our work and related works from two points of view: number of people and interpretation level of position. Many previous works proposed sensing approaches [7–9] that focused on robust and quick estimation of single/multiple human positions (left part of Figure 5.47). Other previous works proposed methods to estimate the crowdedness of environments or human behavior such as walking and visiting an exhibit using position information [10–12] (middle of Figure 5.47). Some other previous works focused on relationships between

---

* This chapter is a modified version of a previously published paper Masahiro Shiomi, Kenta Nohara, Takayuki Kanda, Hiroshi Ishiguro, and Norihiro Hagita, Estimating Group States for Interactive Humanoid Robots, *Proceedings of the 7th IEEE-RAS Annual International Conference on Humanoid Robots (Humanoids 2007)*, 2007, edited to be comprehensive and fit the context of this book.

FIGURE 5.46
Robot simultaneously interacting with many people.

people by using mobile devices [2,13,14] (bottom right of Figure 5.47). These works estimated relationships based on position information but did not estimate group states.

To estimate group states, we focus on two relationships: human relationships between clustered people and position relationships between a robot and clustered people. We expected that such relationships would spatially influence group states. For example, the distance between persons is affected by human relationships. If persons have friendly relationships or shared purposes, the distance between them will be small. Moreover, position relationships between a robot and persons are essential elements to recognize the situation surrounding the robot. For example, in an orderly state, the distances between clustered peoples are small when many people together are listening to a robot, as shown in Figure 5.46. If the surrounding environment is unsuitable for a robot's task, it can change behavior to create a suitable state. Such behavior is needed for an interactive humanoid robot to mingle with multiple people. Therefore, we estimate group states by considering the clustering of people based on human relationships (upper right of Figure 5.47).

In this research, we propose a method that estimates a suitable group state by providing information of multiple people to an interactive humanoid robot. Suitable group states are defined by two coders. The proposed method distinguishes between orderly and disorderly group states using position relationship between a robot, people, and a Support Vector Machine (SVM).

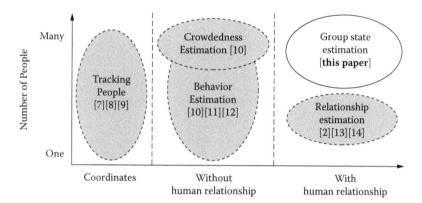

FIGURE 5.47

Comparisons between our work and related works. (a-1) People spread around robot. (a-2) People standing in front of robot. (a-3) People standing in front of robot. (b-1) Small scattered group. (b-2) People stand behind robot. (b-3) People lined up before robot.

We investigate the performance of the proposed method using group state data gathered from a real environment in which a robot simultaneously interacts with multiple people.

5.6.2 Estimating a Group State

In this chapter, we estimate "orderly" and "disorderly" states when a robot is providing information to multiple people. An information-providing task is basic for communication robots that interact with people and perform in real environments. Past works reported the effectiveness of providing information tasks in real environments [2–5].

However, since the definitions of "orderly" and "disorderly" are subjective, we defined an "orderly" state as a suitable situation for an information-providing task. Two coders distinguished group states based on our definitions. We define orderly and disorderly as follows:

Orderly: when multiple (more than five) people surround and face a robot, and their shape seems suitable for receiving information from a robot (upper part of Figure 5.48).

Disorderly: when multiple (more than five) people face a robot and are not surrounding a robot, or their shape does not seem suitable for receiving information from a robot (lower part of Figure 5.48).

We focused on a cluster of people based on distance between people from estimated positions by environmental sensors. This approach enables us to

(a-1)

(a-2)

(a-3)

FIGURE 5.48
Example scenes: (a) orderly and(b) disorderly.

(b-1)

(b-2)

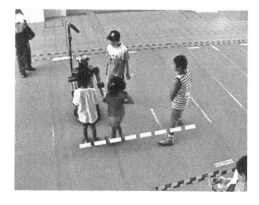

(b-3)

FIGURE 5.48 (continued)

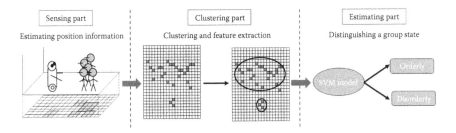

FIGURE 5.49
Outline of developed system with proposed method.

estimate the shape and position relationship of clustered people who have relationships. From this point of view, our proposed method consists of three parts: sensing, estimating, and clustering. Figure 5.49 shows an outline of the proposed method for distinguishing orderly and disorderly group states. In this section, we describe the details of each part.

5.6.2.1 Sensing Part

Here we detected the interaction of people's positions with floor sensors because they can collect high-resolution data, are occlusion free, robust for changes in lighting conditions, and can detect pressure from a robot or people. Therefore, as shown in Figure 5.46, floor sensors estimate people's positions better in crowded situations than such sensors as ceiling cameras and laser range finders.

Our floor sensors are 500 [mm^2] with a resolution of 100 [mm^2]. Their output is 1 or 0; the floor sensor is either detecting pressure or it is not. Floor sensors are connected with each other through an RS-232C interface at a sampling frequency of 5 Hz.

Figure 5.50a shows a floor sensor, and Figure 5.50b shows interaction between a robot and people (upper part) and outputs from the floor sensors (lower part). Black squares indicate reactive points at which a sensor detected pressure from a robot or people.

5.6.2.2 Clustering Part

In the clustering part, the system applies a clustering method to the floor sensor data, splits the clusters based on the distance between two persons (see the proxemics theory of E. T. Hall [15]), and extracts the features of each clustered group. In this section, we describe each component of the clustering part.

5.6.2.2.1 Clustering Method

We can estimate the information of a group's shape by extracting such cluster features as standard deviation and average distance between a robot

(a)

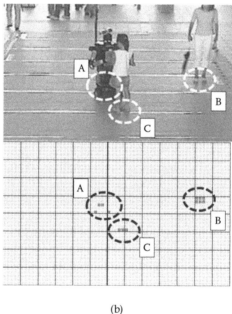

(b)

FIGURE 5.50
Floor sensors. (a) Image of floor sensor. (b) Multiple people on floor sensors.

and a cluster. For example, when people fan out from a robot, standard deviation decreases. When people stand in line near a robot, standard deviation increases.

We applied a nearest-neighbor method for clustering floor sensor data to classify the neighboring reactive points. A nearest-neighbor method has two merits for estimating group states. One, it is hierarchical; it can decide the

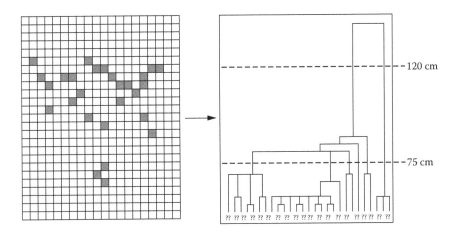

FIGURE 5.51
Cluster tree diagram from floor sensor data.

number of clusters based on cluster distance. But a nonhierarchical method must decide the number of clusters beforehand. In human–robot interaction, deciding the number of groups beforehand is difficult.

Second, the nearest-neighbor method can make oblong-shaped clusters because cluster distance is the distance between the smallest elements of each cluster. When multiple people fan out from a robot, the shape becomes oblong. Therefore, we used the nearest-neighbor method for extracting the shape features of the group interacting with a robot.

Cluster tree diagrams are made by applying the nearest-neighbor method for clustering floor sensors data, as shown Figure 5.51. The vertical axis represents the distance between clusters. Cluster tree diagrams are used for splitting clusters.

5.6.2.2.2 Splitting Clusters

The system splits clusters based on the distance between them because it assumes that adjacent people comprise one group. E. T. Hall reports four groups of distance between people that change based on the relationship, concluding that friends or families maintain a personal distance of approximately 75 [cm] to 120 [cm] (close phase of personal space).

$$psuedo\ t^2 = (\frac{W(R)-W(P)-W(Q)}{W(P)+W(Q)})$$
$$\times (N(P)+N(Q)-1)$$

(5.21)

From this point of view, the system does not split clusters under distances less than 75 [cm], but it always splits clusters with distances of more than

120 [cm]. If the cluster distance is between 75 and 120 [cm], the system used pseudo t² [16] for splitting, which means the separated rate between clusters and indicates the cluster number. In the case of Figure 5.51, the system calculated pseudo t² for splitting because some cluster distances exist between 75 and 120 [cm].

The following equation represents the pseudo t² of cluster R, which combines clusters P and Q. $W(c)$ represents the sum of the distance between the gravity point of cluster c and each of its elements. $N(c)$ represents the number of elements of cluster c. The system calculates pseudo t² in each cluster number. When the positive change of pseudo t² is at maximum, the system uses the number to split the cluster.

5.6.2.2.3 Feature Extraction

The system extracts feature vectors using clustered floor sensor data and robot position information. These feature vectors are used for the estimation part of the proposed method. The following are the extracted feature vectors:

Number of reactive points of floor sensors

Number of clusters

Average distance between a robot and each cluster element

Standard deviation of average distance

Degree between the robot direction and the gravity point of a cluster

The number of reactive points of floor sensors and clusters indicates the degree of congestion around the robot. Average distance, standard deviation, and degree, which represent the position relationships between the robot and each group, were calculated for the three clusters in the order of the number of reactive points. Therefore, the system extracted 11 feature vectors from the clustered floor sensor data.

Figure 5.52 illustrates the clustered floor sensor data when people fan out from a robot. Equation (5.22) represents the average distance between a robot and each element of cluster A. N represents the number of clusters A. $Dist (A_i)$ represents the distance between a robot and reactive point "A_i". Equation (5.23) represents the standard deviation of the average distance between a robot and each element of cluster A. The degree between the front of the robot and a cluster's gravity point is shown in Figure 5.52.

$$\mu_A = \sum_{i=1}^{N} Dist(A_i) \div N \tag{5.22}$$

$$\sigma_A = \sqrt{\left(\sum_{i=1}^{N} (\mu_A - Dist(A_i))^2 \div N \right)} \tag{5.23}$$

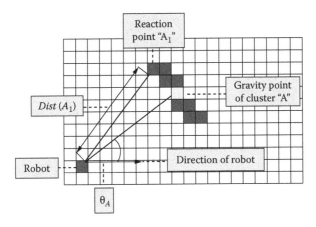

FIGURE 5.52
Extracted feature vectors from floor sensor data.

5.6.2.3 Estimating Part

In the estimating part, the system distinguishes group states (orderly or disorderly) from clustered floor sensor data. For estimating, the proposed method uses an SVM model, which is a representative 2-class classification method [17], because it is generally an efficient learner of large input spaces with a small number of samples.

In this research, we used extracted feature vectors from clustered floor sensor data to construct an SVM model, because for constructing such a model, we must prepare sets of labeled training data that have feature vectors for distinguishing group states.

5.6.3 Experiment

We gathered data from a field experiment to evaluate the effectiveness of our proposed method by installing a humanoid robot named "Robovie" [18], floor sensors, and cameras at a train station to gather group state data. In this section, we describe the details of the evaluation experiment.

5.6.3.1 Gathering Data for Evaluation

We conducted a two-week field trial to gather data of group states at a terminal station of a railway line that connects residential districts with the city center. The station users are mainly commuters, students, and families visiting the robot on weekends. Users could freely interact with the robot. There were four to seven trains per hour.

Figure 5.53 shows the experiment's environment. Most users go down the stairs from the platform after exiting the train. We set the robot and the sensors in front of the right stairway. The robot provided such services as route

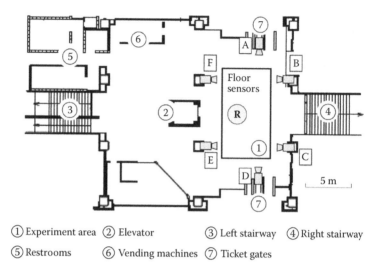

① Experiment area ② Elevator   ③ Left stairway ④ Right stairway

⑤ Restrooms   ⑥ Vending machines ⑦ Ticket gates

FIGURE 5.53
Settings of pre-experimental environment for gathering data

guide and child-like interaction. We set 128 floor sensors at the center of a 4×8 [m] floor area around which the robot was placed and moved. In addition, we recorded the images from six cameras to classify scenes into two classes by the two coders.

In the experiment, we gathered a lot of position data that express the interaction scenes between a robot and a group of people, although the experiment was performed at a train station. We believe that ordinary people treated the robot as an exhibit because robots are too novel for them. In fact, more than 1000 people interacted with our robot in the experiment. In addition, we observed scenes where a group surrounded the robot, as shown in Figure 5.48a and 5.48b. We gathered 152 scenes where more than five people stood still around the robot for more than five seconds.

5.6.3.2 Making an SVM Model Using Gathered Data

The coders classified the gathered scenes using recorded images. As a result, 72 disorderly scenes were observed (1st week: 36, 2nd week: 36) and 36 orderly scenes were observed (1st week: 18, 2nd week: 18). Namely, 108 scenes were consistent, and 44 scenes were not consistent.

The kappa statistic is used to subjectively investigate the corresponding ratio between multiple people [19]. If the κ statistic equals 0.40, it indicates that the subjective of multiple people is middling consistent. The kappa statistic between the two coders' subjective evaluations was 0.49, indicating that their evaluations are considerably consistent.

For making training data, we used 36 disorderly and 18 orderly scenes gathered from the 1st week's field trial. For the disorderly scenes, the system extracted feature vectors using clustered floor sensor data at two seconds; the system extracts feature vectors five times from one scene. For the orderly scenes, the system extracts feature vectors using clustered floor sensor data at four seconds; the system extracts feature vectors twenty times from one scene. In addition, we made dummy scenes based on each scene by reversing the X-axis floor sensor data. Thus, we made three dummy scenes from one scene and prepared 720 training data for the orderly and disorderly scenes.

The amount of test data for evaluations was also 720 for each scene; these data were made using the data gathered from the second week's field trial. Thus, we made an SVM model using data gathered from the field trial in a real environment and evaluated the performance of the SVM model for unknown data.

5.6.3.3 *Evaluation of Proposed Method*

Evaluation was performed using the LIBSVM library [20]. The SVM and kernel parameters were determined by attached tools to search for effective parameters [20]. To evaluate the proposed method, we compared the accuracy between the proposed method (11 feature vectors), raw data (3200 feature vectors), Haar wavelet transform (3200 feature vectors), and Daubechie wavelet transform (3200 feature vectors), because raw data and wavelet transforms with SVM are often used for pattern recognition.

Figure 5.54 shows the accuracy average of the SVM model in the test data. In the proposed method, estimation accuracy is 80.4% (orderly group state is 81.8% and 78.9% for the disorderly group state). These results indicate that the performance of the SVM model is 80.4%, even though unknown data were used.

By using the proposed method, the accuracy of the SVM model was better than other SVM models (Raw data is 50%, Haar wavelet is 58.1%,

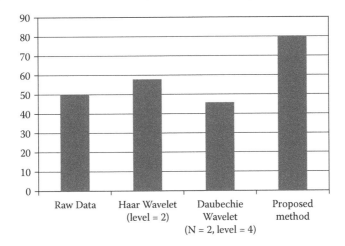

FIGURE 5.54
Average accuracy of estimating group states.

and Daubechie wavelet is 46.1%), indicating that the proposed method outperformed other feature extraction methods. This result demonstrates the improvement in group state estimation enabled by our approach that uses clustering based on the proxemics theory. A group is constructed from people, so it is reasonable to suppose that human relationships will affect group states. The proposed method extracted effective feature vectors to estimate the group states by using the proxemics theory. In addition, eliminating useless feature vectors also improved accuracy.

5.6.4 Discussion

5.6.4.1 Contributions for Human–Robot Interaction

Based on the proposed method, the SVM model can distinguish orderly and disorderly scenes with 80.4% accuracy. These values are better than using other feature extraction methods such as wavelet transform functions. We believe that the proposed method enables communication robots to estimate whether the group state is suitable for providing information in other daily environments because we gathered many interaction scenes between a robot and group of people, even though the experiment was performed at a train station. In addition, the proposed method can easily be applied to other systems that include position estimation functions using other kinds of sensors. Therefore, robots can influence a group of people to create a suitable situation for its task using this method.

Our past work proposed and evaluated a behavior design for interaction between a robot and a group in such crowded situations [21]. The robot's system also included a human operator who controlled part of the recognition function for interaction in a group. We believe that a robot can autonomously interact with a group in crowded situations by implementing the proposed method and behavior design.

In addition, we expect to apply the proposed method to estimate other kinds of suitable group states for other robot tasks, such as guiding and playing, because it can distinguish scenes defined by multiple coders. Moreover, the proposed method can be applied to situations without communication robots such as surrounding, standing, or gathering people around an exhibit or ticket gate. Therefore, we can make an SVM model that estimates suitable group states for any tasks or situations using labeled scenes by multiple coders and floor sensor data.

5.6.4.2 Performance Improvement Approach

From the results of experiments, we found three kinds of problems that caused error. In this section, we discuss the details of the three problems and three approaches for improving our proposed method's performance.

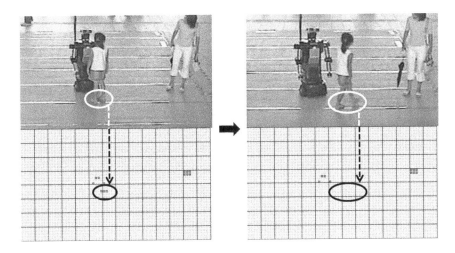

FIGURE 5.55
Example scenes when floor sensors "lost" a child.

5.6.4.2.1 Sensing Problems

Figure 5.55 shows scenes in which the floor sensors could not detect a child who slightly changed position. The main reason apparently involved the weight of the people. Sometimes such situations were observed when children were directly on the floor sensors. Although the system was robust to occlusion problems, it could not correctly estimate the positions of interacting people in such situations.

Combining other kinds of sensors such as multiple cameras and laser range finders might estimate position information more correctly. Other kinds of information such as the directions of faces and people's behavior are also effective for estimating group states in detail, if the system correctly estimates such information. In addition, using other types of floor sensors that can output analog values based on pressure will help solve such sensing problems.

5.6.4.2.2 Training Data Problems

The experiment results show that the performance of estimating disorderly scenes with test data was much lower than using training data. We expected a wide variety of disorderly scenes, so the training data did not include some kinds of disorderly scenes. To improve the performance of the proposed method, we must prepare more training data of orderly and disorderly scenes for making an SVM model.

5.6.4.2.3 Clustering Problems

In the clustering part of the proposed method, the system applied a cluster method to the reactive points of the floor sensor data. We expect the system to estimate the number of people and the group's shape more correctly if it can track each person on the floor sensors. Using such information will improve the performance of the proposed method. Person-tracking functions with floor sensors have already been proposed [8,22], so we can easily apply them to our proposed method. In addition, we will be able to extract more feature vectors with floor sensors such as number of people and staying time.

5.6.5 Conclusion

In this chapter, we proposed a method for estimating whether group states are orderly or disorderly by floor sensor data and a clustering method. To estimate the group states, we focused on position relationships between people who have a relationship, such as families. Floor sensors detect position information of people around the robot. The clustering method consists of the nearest-neighbor method based on the proxemics theory [15], which is the distance between two persons and pseudo t^2.

To investigate the proposed method's performance, we gathered floor sensor data in a field trial at a train station. Using the gathered data, the proposed method correctly estimated the group states at an 80.4% rate; our proposed clustering method based on the proxemics theory outperformed other feature extraction methods such as wavelet transform. The proposed method enables communication robots to estimate whether the group state is orderly or disorderly when interacting with many people.

In future work, we will continue to improve the proposed method and use it to develop a robot that interacts with multiple people in real environments.

Acknowledgments

We wish to thank the staff of the Kinki Nippon Railway Co. Ltd. for their kind cooperation. We also wish to thank the following ATR members for their helpful suggestions and cooperation: Satoshi Koizumi and Daisuke Sakamoto. This research was supported by the Ministry of Internal Affairs and Communications of Japan.

References

1. Asoh, H., Hayamizu, S., Hara, I., Motomura, Y., Akaho, S., and Matsui, T. Socially embedded learning of the office-conversant mobile robot Jijo-2, *Int. Joint Conf. on Artificial Intelligence*, pp. 880–885, 1997.
2. Kanda, T., Hirano, T., Eaton, D., and Ishiguro, H. Interactive robots as social partners and peer tutors for children: A field trial, *Human Computer Interaction*, Vol. 19, No. 1–2, pp. 61–84, 2004.

3. Siegwart, R., et al. Robox at Expo. 02: A large scale installation of personal robots, *Robotics and Autonomous Systems*, 42, pp. 203–222, 2003.
4. Shiomi, M., Kanda, T., Ishiguro, H., and Hagita, N. Interactive humanoid robots for a science museum, *1st Annual Conference on Human-Robot Interaction*, 2006.
5. Tasaki, T., Matsumoto, S., Ohba, H., Toda, M., Komatani, K., Ogata, T., and Okuno, H. G. Distance based dynamic interaction of humanoid robot with multiple people, *Proc. of Eighteenth International Conference on Industrial and Engineering Applications of Artificial Intelligence and Expert Systems*, pp. 111–120, 2005.
6. Kendon, A. Spatial organization in social encounters: the F-formation system, A. Kendon, Ed., *Conducting Interaction: Patterns of Behavior in Focused Encounters*, Cambridge University Press, 1990.
7. Cui, J., Zha, H., Zhao, H., and Shibasaki, R. Laser-based interacting people tracking using multi-level observations, *Proc. 2006 IEEE/RSJ Int. Conf. on Intelligent Robots and Systems*, 2006.
8. Murakita, T., Ikeda, T., and Ishiguro, H. Human tracking using floor sensors based on the Markov chain Monte Carlo method, *International Conference on Pattern Recognition*, pp. 917–920, 2004.
9. Shiomi, M., Kanda, T., Kogure, K., Ishiguro, H., and Hagita, N. Position estimation from multiple RFID tag readers, *The 2nd International Conference on Ubiquitous Robots and Ambient Intelligence (URAmI2005)*, 2005.
10. MacDorman, K. F., Nobuta, H., Ikeda, T., Koizumi, S., and Ishiguro, H. A memory-based distributed vision system that employs a form of attention to recognize group activity at a subway station, *IEEE/RSJ International Conference on Intelligent Robots and Systems (IROS)*, pp. 571–576, Sep. 2004.
11. Liao, L., Fox, D., and Kautz, H. Location-based activity recognition using relational Markov networks, *Int. Joint Conf. on Artificial Intelligence (IJCAI-05)*, 2005.
12. Kanda, T., Shiomi, M., Perrin, L., Nomura, T., Ishiguro, H., and Hagita, N. Analysis of people trajectories with ubiquitous sensors in a science museum, *IEEE International Conference on Robotics and Automation (ICRA2007)*, 2007 (to appear).
13. Eagle, N., and Pentland, A. Reality mining: Sensing complex social systems, *Personal and Ubiquitous Computing*, online first, Nov., 2005.
14. Choudhury, T., and Pentland, A. Modeling face-to-face communication using the sociometer, *Int. Conf. Ubiquitous Computing (Ubicomp2003)*, 2003.
15. Hall, E. T. *The Hidden Dimension*. Anchor Books, 1990.
16. Milligan, G., and Cooper, M. Methodology review: Clustering methods, *Applied Psychological Measurement*, Vol. 11, pp. 329–354, 1987.
17. Vapnik, V. *The Nature of Statistical Learning Theory*, Springer (1995).
18. Ishiguro, H., Ono, T., Imai, M., Maeda, T., Kanda, T., and Nakatsu, R. Robovie: an interactive humanoid robot, *Int. J. Industrial Robot*, Vol. 28, No. 6, pp. 498–503, 2001.
19. Carletta, J. Assessing agreement on classification tasks: the kappa statistic, *Computational Linguistics*, Vol. 22(2), 249–254, 1996
20. Chang, C. C., and Lin, C. J. LIBSVM: Introduction and Benchmarks, http://www.csie.ntu.edu.tw/cjlin/libsvm.
21. Shiomi, M., Kanda, T., Koizumi, S., Ishiguro, H., and Hagita, N. Group attention control for communication robots, *ACM 2nd Annual Conference on Human-Robot Interaction (HRI2007)*, 2007.
22. Silve, G. C., Ishikawa, T., Yamasaki, T., and Aizawa, K. Person tracking and multicamera video retrieval using floor sensors in a ubiquitous environment, *Int. Conf. on Image and Video Retrieval*, pp. 297–306, 2005.

6

Shared Autonomy and Teleoperation

6.1 Introduction

How will social robots be deployed in our daily lives? It would be great if fully autonomous robots work in the real world as designed; however, this would need a lot of technologies that are not available today. Robots have difficulty in recognizing people's voices in noisy environments. Robots do not know about people and environments, and thus there is a big knowledge gap, perhaps to be filled by some learning technique in the future. Do we need to wait for such techniques to become available?

This situation would be somewhat similar across many robotics fields, such as space exploration, search and rescue, medicine, and the military domain. The common solution is not to wait for full autonomy, but to put humans into the loop to supplement immature technologies. This is called *shared autonomy*. As robots are networked, it would be natural to assume that humans are connected to a robotic system. Previous studies and applications have demonstrated cases in which small contributions from human operators or supervisors would make semi-autonomous robotic systems feasible for immediate deployment. A simple example is an exploration robot, for example, Mars Rover, which is frequently teleoperated from operators located on Earth, yet the robot has some autonomy such as collision avoidance [1].

Plenty of research has been done on shared autonomy. One of the important questions is "when to control the robot." If a robot were to occupy an operator's attention, teleoperation would be less feasible for social robots. One idea is to adjust the level of autonomy, that is, *adjustable autonomy* [2] or *sliding autonomy* [3], so that human operators are only involved in the loop when needed. Another aspect is *situation awareness* [4], the operator's perception of the situation around a robot. There are some studies on designing user interface to provide better situation awareness; for instance, it has been shown that an interface with an integrated view works better, as an operator finds it difficult to pay attention to separate windows [5,6]. Research has been conducted on the analysis of *fan-out*, the overall performance of the system [7].

There are only a few studies on teleoperation of social robots. Kuzuoka et al. has shown the existence of a "dual-ecology" aspect: an interface that provides better situation awareness to an operator is not necessarily a good interface for a user who is interacting with the robot [8]. In a telepresence use of an android robot, automation is conducted for low-level behavior to make the robot's behavior lifelike [9].

This chapter introduces two instances of shared autonomy study in social robotics. In the first study, we report a rule-based semi-autonomous system, and show how such semi-autonomy could contribute to improving the performance of robotic systems. Yet, one of the downsides was the fact that human operators sometimes failed to respond although autonomy failed to handle the situation. In the second study, we report a technique to divide operators' attention among multiple robots. The key concept is a technique to delay social interaction so that operators' attention is allocated in a timely manner to a robot that needs assistance.

We consider shared autonomy to be the necessary approach toward deployment. In Chapter 2, Section 2.2, we show a case in which teleoperation approach enabled deployment as well as data collection to increase "covered situation" while the robot is in real use. In one of our later studies, we report a way to formalize *fan-out* in social robot teleoperation and use such formalization in the deployment process [10]. There are apparently many open issues. *Situation awareness* would require many more studies. For instance, "dual-ecology" aspects are seen in the temporal aspects as well [11]. When social robots move around, we have observed that interaction between robots and an operator is much more complicated [12].

References

1. Fong, T., and Thorpe, C. (2001). Vehicle teleoperation interfaces. *Autonomous Robots, 11*(1), 9–18.
2. Goodrich, M. A., Olsen, D. R., Crandall, J. W., and Palmer, T. J. (2001). *Experiments in Adjustable Autonomy*. Paper presented at the IEEE Int. Conf. on Systems, Man, and Cybernetics (SMC2001).
3. Fong, T., Thorpe, C., and Baur, C. (2001). *Collaboration, Dialogue, Human-Robot Interaction*. Paper presented at the International Symposium of Robotics Research.
4. Endsley, M. R. (1988). *Design and Evaluation for Situation Awareness Enhancement*. Proc. of the Human Factors Society 32nd Annu. Meet. 97–101.
5. Nielsen, C. W., Goodrich, M. A., Member, S., and Ricks, R. W. (2007). Ecological interfaces for improving mobile robot teleoperation. *IEEE Transactions on Robotics, 23*(5), 927–941.
6. Yanco, H. A., Drury, J. L., and Scholtz, J. (2004). Beyond usability evaluation: Analysis of human-robot interaction at a major robotics competition. *Human-Computer Interaction, 19*(1&2), 117–149.

7. Crandall, J. W., and Cummings, M. L. (2007). *Developing Performance Metrics for the Supervisory Control of Multiple Robots.* Paper presented at the ACM/IEEE Int. Conf. on Human-Robot Interaction (HRI2007).
8. Kuzuoka, H., Yamazaki, K., Yamazaki, A., Kosaka, J. I., Suga, Y., and Heath, C. (2004). *Dual Ecologies of Robot as Communication Media: Thoughts on Coordinating Orientations and Projectability.* Paper presented at the ACM Conference on Human Factors in Computing Systems (CHI2004).
9. Sakamoto, D., Kanda, T., Ono, T., Ishiguro, H., and Hagita, N. (2007). *Android as a Telecommunication Medium with a Human-like Presence.* Paper presented at the ACM/IEEE Int. Conf. on Human-Robot Interaction (HRI2007).
10. Zheng, K., Glas, D. F., Kanda, T., Ishiguro, H., and Hagita, N. (2011). *How Many Social Robots Can One Operator Control?* Paper presented at the ACM/IEEE 6th Annual Conference on Human-Robot Interaction (HRI 2011).
11. Glas, D. F., Kanda, T., Ishiguro, H., and Hagita, N. (2011). Temporal Awareness in Teleoperation of Conversational Robots. *IEEE Transactions on Systems, Man, and Cybernetics—Part A: Systems and Humans.*
12. Glas, D. F., Kanda, T., Ishiguro, H., and Hagita, N. (2009). *Field Trial for Simultaneous Teleoperation of Mobile Social Robots.* Paper presented at the ACM/IEEE Int. Conf. on Human-Robot Interaction (HRI2009).

6.2 A Semi-Autonomous Social Robot That Asks Help from a Human Operator

Masahiro Shiomi,[1] Daisuke Sakamoto,[2] Takayuki Kanda,[1]
Carlos Toshinori Ishi,[1] Hiroshi Ishiguro,[1,3] and Norihiro Hagita[1]

[1]*ATR IRC Laboratories,* [2]*The University of Tokyo,* [3]*Osaka University*

ABSTRACT

The capabilities of autonomous robots are certainly limited. To what extent can a human operator improve these capabilities? This study reports our development of a semi-autonomous social robot. The robot is developed with a "network robot system" approach. That is, a humanoid robot is connected with sensors embedded in the environment over network. There is a rule-based mechanism to ask an operator to help the robot. The robot is tested in a train station for its role, which is to greet travelers and give them route guidance. The result demonstrated that ratio of success in providing route guidance was much improved. A fully autonomous robot only provides 29.9% successful guidance; the semi-autonomous robot was successful in providing 68.1% successful guidance, while a human operator only operated 25% of the time.

6.2.1 Introduction*

Although robots would be networked and thus can get information about people's behavior from sensors in the environments, yet robot's autonomy can deal with only limited situations. There would be a number of awkward situations or errors caused by autonomy. In particular, robots cannot yet communicate verbally. Speech recognition of colloquial utterances in noisy environments is still difficult, particularly when the microphone is distant from the speaker; thus, environmental noise would have to be seriously considered.

Moreover, there is a basic difficulty in the development process. It is hard to anticipate people's behavior in real environments. Thus, to know them, we would need to put robots in real situations. However, it is difficult to install immature robots. It would invite complaints, and would result in rejection of the robot.

In this chapter, we report a study on the development of a semi-autonomous robot to cope with the above problem. As robots are expected to be networked, it would be natural to assume that human developers, supervisors, and operators would have access to them over a network. People have used the Wizard of Oz (WOZ) method [1,2], though in previous human–robot interaction studies, little autonomy was involved. We have observed that it is difficult for a human operator to be responsible for the full functionality of a robot. Our semi-autonomous approach has a similar thrust to that of the WOZ method: use of a human operator with a prototype system to gather realistic data from users. Moreover, we wonder whether a robot would be able to ask assistance from a person in case a problem is detected. This would enable a robot to operate in a real field with an acceptable performance from the beginning, while enabling human developers to improve the robot's capability based on the interaction with the robot.

6.2.2 Semi-Autonomous Robot System

6.2.2.1 Overview

Figure 6.1 shows an overview of the system. The system is designed with the idea of having three layers: *reflective, behavioral,* and *reactive.* This design is informed by D. Norman's human model (p. 28 in [3]). We consider that *reactive* things (innate actions embedded to humans) can be addressed by autonomy. Moreover, patterned behavior for specific tasks and situations learned from experience, so-called behavior, can be automated too. In contrast, we consider that it is difficult for an autonomous robot to address reflective things,

---

* This chapter is a modified version of a previously published paper Shiomi, M., Sakamoto, D., Kanda, T., Ishi, C. T., Ishiguro, H., and Hagita, N. (2008). A Semi-autonomous Communication Robot—A Field Trial at a Train Station. Paper presented at the *ACM/IEEE Int. Conf. on Human-Robot Interaction (HRI2008),* edited to be comprehensive and fit the context of this book.

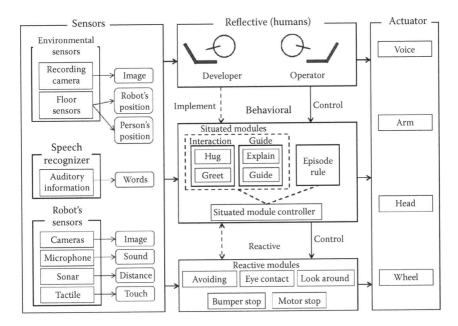

FIGURE 6.1
System overview.

for example, working with a situation it encounters for the first time, and scenes that requires in-depth thought. Here, the system has a mechanism that autonomously asks the human operator to help the robot.

For the system, we used Robovie, an interactive humanoid robot characterized by its human-like physical expressions and its various sensors. There are couple of functionalities implemented that are explained in other sections. Robovie is capable of generating a pointing gesture (Chapter 4, Section 4.2) for direction giving (Chapter 4, Section 4.3). Floor sensors (Chapter 5, Section 5.4) are connected over the network. In addition, a speech recognition software, ATRASR [4], is used with a noise filter [5].

6.2.2.2 Reactive Layer

The conceptual purpose of the reactive layer is to achieve safe interaction and lifelike behavior. In order to provide safe interaction, the robot's locomotion and motors stop when an object contacts the robot's bumper or an overload of any motor is detected. There are reactive behaviors for exhibiting the robot's lifelikeness, such as eye contact and gazing at touched body parts. These reactive behaviors were prepared for a general environment. In other words, we only implemented simple mechanisms in the reactive layer that do not work incorrectly and do not require software updating.

6.2.2.3 Behavioral Layer

The conceptual purpose of the behavioral layer is to realize task-dependent and environment-dependent behaviors. We used our behavior-based architecture [6] for this purpose. In this architecture, the behavioral layer consists of situated modules, a situated module controller, and episode rules. Situated modules realize the robot's interactive behavior with situation-dependent sensory data processing for recognizing reactions from humans. Because each module works in a particular situation, a developer can easily implement situated modules by taking into account only the particular limited situation. A situated module is implemented by coupling communicative sensory-motor units with other directly supplementing sensory-motor units such as utterances and gestures. Episode rules describe the rule of state transition among the situated modules. With the behavioral layer, the robot can autonomously interact with people. We have implemented two types of situated modules: guidance behavior and greeting and free-play behaviors.

6.2.2.3.1 Situated Module for Greeting and Free-Play Interaction

First, the robot approaches a person detected by the floor sensors. It initiates the interaction by greeting and offering handshaking. If the person requests route guidance by saying, "would you tell me a route?" it immediately starts guidance behaviors (if the utterance is correctly recognized). Adults often do this. The robot is also capable of free-play behavior, which is popular with children. The robot sometimes triggers tactile interaction with a child by saying, "Let's play, touch me"; once the child reacts, it continues performing free-play behaviors as long as the child responds. It initiates some brief chatter, such as "where are you from?" It also offers a hug, which is much liked by children. At the end of the interaction, the robot exhibits bye-bye behavior.

The robot also offers information around the station as chatter, such as "there is a new shopping center built close to the station." After it exhibits several free-play behaviors, it initiates guiding behavior.

6.2.2.3.2 Situated Module for Guiding

The robot was able to guide visitors to 38 nearby places. The robot offers route guidance by asking, "Where are you going?" If an interacting person responds, it starts to describe how to go to the place. For example, when guiding a visitor to the bus stop, the robot points toward an exit and says, "Please go out this exit and turn right. You can see the bus stop immediately." When the robot explains the route, it utilizes a pointing gesture as well as reference terms such as this and that. In Japanese, there are three types of reference terms associated with positional relationships among two interacting persons and the object they are talking about. We installed a model for these reference terms and pointing gestures, called the "three-layer attention-drawing model" [7]. Thus, the robot autonomously generates a behavior to guide visitors to these destinations using appropriately chosen reference terms and gestures. In addition, it has a map for these locations. If the interacting person cannot directly see the destination, such as a place outside of the station, the robot points to a visible place, such as the exit, and verbally supplements the remaining directions.

6.2.2.3.3 Episode Rule—Transition Rules for Situated Modules

The relationships among behaviors are implemented as rules governing an execution order (named "episode rules") to maintain a consistent context for communication. The basic structure of the episode rule consists of previous behaviors (e.g., "greeting" behaviors finished as "successful") and the next behaviors (e.g., "offering route-guidance" behavior). The situated module controller selects a situated module based on episode rules. We have implemented 1311 Episode Rules. As described above, the episode rules were designed to realize these behaviors: (1) approaching a visitor, (2) extending a greeting, (3) offering chatter, information around the station, and free-play, (4) offering route guidance, and (5) saying goodbye. In addition, event-driven transition was described so that when a passenger initiates route-guidance conversation, the robot begins to offer route guidance.

6.2.2.3.4 Example of Transition

Figure 6.2 shows one example of episode rules and scenes of an interaction between a robot and a person. In this case, the floor sensors detect the person's

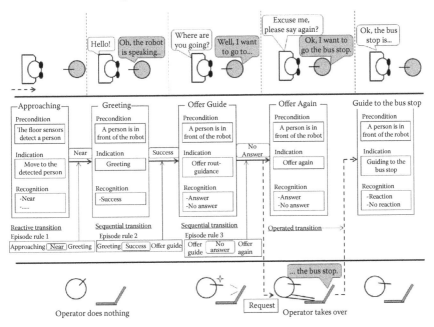

FIGURE 6.2
An illustration of episode rules and scenes of an interaction with an operator.

position, and then the robot approaches the person by the situated module of "Approaching." This causes a reactive transition ruled by episode rule 1. The robot greets the person by executing the situated module "Greeting." After "Greeting," the robot provides route guidance by executing "OfferGuide." In this example, since the visitor does not respond or the speech recognizer fails to pick up what the visitor said, the "OfferGuide" results in "No answer," and then the robot asks the visitor again using "OfferAgain." At the same time, the operator-requesting mechanism (described below in III-D.1) fires so that the operator is asked to take over control. The visitor might answer the robot with a response such as "I'd like to go to the bus stop," which is heard by the operator. As the result of the operator's control, "Guide to the bus stop" is finally selected.

6.2.2.4 Reflective Layer

The conceptual purpose of the reflective layer is to involve humans in the loop so that the system as a whole can process natural language, think deeply, and improve itself with human support. The robot is able to autonomously operate without the reflective layer. In addition, using the reflective layer, the system requests help from a human operator as well as accumulates the data needed by a human developer for making further improvements.

The key mechanism is the algorithm to request the operator's help. The system detects when the robot cannot handle a situation by itself and then requests a human operator to take over its control. The system has two ways to deal with detecting errors: *report from subsystem* and *monitoring of interaction pattern*.

6.2.2.4.1 Report from Subsystem

Each subsystem individually reports a problem to request the assistance of an operator. For example, in the reactive layer, the robot stops its body movements and locomotion when it detects an overload of a motor, which happens when the motor motion is blocked by physical collision with people or other objects. When such a report is made, the robot system requests operator support. The speech-recognition module reports interaction-level errors. It monitors several negative words such as "I don't understand," "That's not right," and "That isn't what I asked." Any of these phrases indicate a problem at the level of human–robot interaction.

6.2.2.4.2 Monitoring of Interaction Pattern

The system monitors pattern of occurred interaction to detect problems. The architecture allows reactive transition of behavior, though according to the design we expect some patterns not to occur if things go well. The system monitors behavior transitions. There is a pattern of problematic situations defined in advance. The robot system requests the operator's help when a transition pattern matches such a predefined pattern of defining problematic situation. To be specific, three patterns are prepared:

(1) Failure in offering successful route guidance: This is detected as a pattern in which the route-guiding behavior results in "no-answer" twice (as shown in Figure 6.2). This rule is designed to match two situations: (a) an interacting person does not speak because he or she is not interested in the route guidance but would be interested in something else, or (b) the speech recognition module fails. In either case, the operator's help is needed.

(2) Failure in initiating route guidance: This is detected as a pattern in which the robot continues interactive behavior over ten times without performing the route-guiding behavior. The robot is designed to start the route-guiding behavior within a few interactive behaviors if people follow ordinary flows of interaction. However, people's unexpected behavior would cause a different flow of interaction. For example, some people ignore greeting behavior and instead keep treating the robot like a pet animal. Such persons do not respond to the robot's handshake request even if the robot asks repeatedly, and instead initiate interaction by touching its shoulder.

(3) Offering excessive route guidance: This is detected as a pattern in which the robot continuously offers route guidance three times. The robot has been designed to exhibit route-guidance behavior when it recognizes a spoken request from people, such as "please tell me a route." Thus, the pattern would happen if (a) an interacting person is excessively interested in route-guidance behavior or (b) the robot continuously fails in its route guidance, which results in continuous requests for route guidance.

6.2.3 Field Trial at a Train Station

The developed system was used in the field. We conducted six days of a field trial at the train station. The developed system was placed in the train station. During the field trial, two operating conditions were used in turn: semi-autonomous and completely autonomous.

6.2.3.1 Environment and Settings

1. Environment: The experiment was conducted at a terminal station of a railway line that connects residential districts with the city center. Station users are mainly commuters and students, and families visited the station on a holiday to see the robot. There are four to seven trains every hour. Figure 6.3 shows the experiment's environment. Most users go down the stairs from the platform after they exit a train. We set the robot and the sensors in front of the right stairway (Figure 6.3). We informed visitors that the robot is a test-bed with a

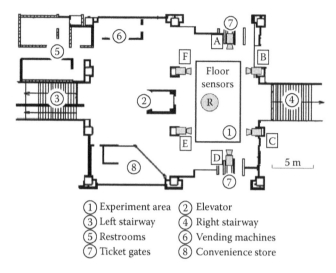

FIGURE 6.3
Station map (1: Experiment area; 2: Elevator; 3: Left stairway; 4: Right stairway; 5: Restrooms; 6: Vending machines; 7: Ticket gates; 8: Convenience store).

route-guidance capability. As shown in Figure 6.3, we placed floor sensors at the center of a floor area of 4 × 8 m. The robot was placed and moved around on the floor sensors.

2. Participants: The users can freely interact with the robot. We asked people who interacted with the robot to fill out a questionnaire after the interaction. Children were asked to fill out questionnaires if they understood the meaning of it.

3. Conditions: The purpose of the autonomous mode is to reveal how much the robot can do without humans' help. We prepared several time slots within a day for each condition in order to keep both conditions the same for fair comparison.

   Autonomous condition: The robot did not use the functions in the reflective layer; thus, the operator did not take over control at all.

   Semi-autonomous condition: The robot uses all implemented layers: reactive, behavioral, and reflective. It usually operates autonomously if there is no visitor around it, where the operator just left the situation as is without taking over the robot's control.

   The operator controlled the robot's behavior only when the operator-requesting mechanism (described in Section 6.3.1.5.1) detected the need for operator help. The operator did not observe the interaction unless the robot asked for help so that we can fairly observe whether this semi-autonomous mechanism works. When an interaction between the robot and the target visitor finished, the operator gave up control, and the system started to work autonomously.

4. Evaluation: The following criteria were used to evaluate the robot's performance:

Success rate of route guidance. This evaluation criterion is directly related to the robot's performance of the route-guidance task. For each visitor who responded to the robot's route-guidance offer or asked the robot about a route, we assume that the robot was successful in giving route guidance if it correctly offered one or more route-guidance directions. Thus, even when a visitor asked for routes to more than one place, one successful guidance was judged a success. The analysis was conducted with the recorded video, and we only analyzed the interaction of the visitors who answered the questionnaire.

Questionnaires for subjective impressions. We asked the visitors who interacted with the robot to answer a questionnaire in which subjects rated items on a scale of 1 to 7, where 7 is the most positive. The following items were used:

Explanation: The degree to which you understood the robot's explanations.

Understanding: The degree to which the robot understood you.

Naturalness: The degree of naturalness of the robot's behavior.

Safety: The degree to which you think the robot is safe.

Success rate of speech recognition. We also evaluated the speech recognizer's performance. This performance is very relevant to the task performance in the autonomous condition. This is calculated as the number of correct answers per number of speech utterances that the robot recognized. Thus, if the robot system fails to detect the speech, such as a voice that is too low, it is not counted. The correct answer is defined as speech where the speech recognizer outputs the recognized word whose meaning matches that of the visitor's speech.

6.2.3.2 *Results*

6.2.3.2.1 *Task Performance*

Figure 6.4 shows the success rate of route guidance in each condition. In Autonomous condition, the success rate is 29.9% among 77 visitors. The main reason of failure was an error in speech recognition. In Semi-autonomous condition, the success rate is 68.1% among 91 visitors. There is still 31.9% failure, which is mainly due to the visitors who immediately stopped interacting with the robot before the operator took over control. In addition, failure in speech

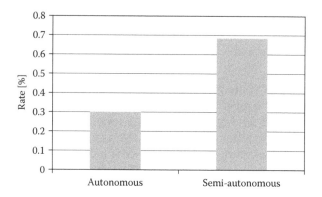

FIGURE 6.4
Success rate of route guidance.

recognition sometimes caused the failure of an operator-request action. If the speech recognizer simply failed to detect the speech or the recognition result was rejected because it did not match the pre-assumed model (this often happened), the operator was successfully requested. The problem was when the speech recognizer picked up a false positive result, which resulted in mistaken guidance while the system could not detect the situation as problematic.

6.2.3.2.2 Subjective Impression

Figure 6.5 shows the results from questionnaires. We conducted a statistical test to verify the differences between the semi-autonomous and autonomous conditions. The analysis of variance (ANOVA) revealed that there are significant differences between the conditions for "Explaining," "Understanding," and "Naturalness" impressions ($p < .01$). This indicates that the subjects rated the semi-autonomous robot more highly than the autonomous robot in these impressions. There is no significant difference in the "Safety" impression.

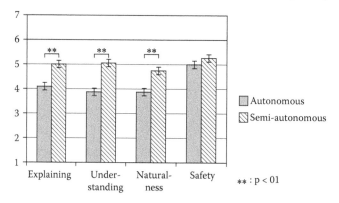

FIGURE 6.5
Subjective impressions.

6.2.3.2.3 *Successful Rate of Speech Recognition*

The system detected 1,571 sentences during the field trials, of which 334 of them were correctly recognized. The success rate was only 21.3%, although in the laboratory it achieved word accuracies of more than 90% with 70 dBA of background noise [8].

There were several reasons for failure in speech recognition: mismatching with prepared language model, missing vocabularies, excessively low volume of speech, excessively loud voice (mainly from a child), and non-constant environmental noise. In the station, the noise level was usually between 65 and 70 dBA, which is not quiet but still not critical for the speech recognizer.

6.2.3.2.4 *Operating Time and Operator-Requesting*

In semi-autonomous condition, the experimental time was 45,900 s, whereas the operating time (the time of the operator controlling the robot) was 11,527 s. Thus, the operator controlled the robot 25% of the experimental time in the semi-autonomous condition.

We also evaluated the performance of the operator-requesting mechanism. Table 6.1 shows this performance. In the table, "Operator requested" represents the interaction with a visitor where the mechanism requests the operator's help, while "Operator NOT requested" represents interaction without the operator's help. "Operator needed" represents the situation where the robot system fails to autonomously offer correct route guidance or does not autonomously offer route guidance. Thus, the upper-left cell "Operator requested" and "Operator needed" indicates an interaction where the robot fails to offer the route guidance at the beginning and then the operator controls the robot to offer the route guidance, which is the ideal situation where the semi-autonomous mechanism works well. In contrast, "Operator NOT requested" and "Operator NOT needed" indicates an interaction where the robot successfully offered route guidance autonomously without the operator's help. The total number of these two types of successful cases was 77, while there were 1 nonnecessary request and 13 failures in requesting. Thus, the operator-requesting mechanism correctly requested the operator's help for 85% (=77/(77+1+13)) of the necessary situations. We believe that this comparison demonstrates good performance in requesting the operator, although the operator was called in the majority of the cases.

TABLE 6.1

Success Rate of the Operator-Requesting Mechanism

	Operator Needed	Operator NOT Needed
Operator requested	72	1
Operator NOT requested	13	5

There is a trade-off between task performance and the operating time. That is, higher operating time will result in better task performance but requires more elaborate control by the operator. In our case, we designed the system to minimize the operator's time; apparently, if the operator engaged in control of the robot from the beginning of the interaction, the task performance was higher.

6.2.4 Conclusion

The study aims to reveal the performance improvement due to having a human operator behind a networked robot system, who sometimes provides support if requested. Thus, the robot is networked not only with sensors but also with human operators. The system autonomously detects a situation in which the operator's help is useful. The field trial at a train station revealed the effectiveness of the developed semi-autonomous system. The operator-requesting mechanism correctly requested the operator's help in 85% of the necessary situations; the robot system successfully guided 68% of the visitors, and improved visitors' subjective impressions. The operator only controlled the interaction 25% of the experiment time.

While the study revealed usefulness of semi-autonomous approach, there are some limitations. First, there are a considerable number of unsuccessful cases in which visitors failed to receive correct route guidance. In the field study, participants rather seemed to want to have fun interacting with the robot; thus, this would be not a problem for the field trial. Nevertheless, this makes it difficult to apply the robot to serve in the route guidance role. It implies that greater accuracy requires an alternative mechanism, maybe by not letting the system autonomously call an operator, as there are always errors expected in autonomy. Second, the operator-requesting mechanism is mostly with hard-coded rules. It would be not easy to find an appropriate rule in other scenarios.

Acknowledgments

We wish to thank the staff of the Kinki Nippon Railway Co. Ltd. for their kind cooperation. We also wish to thank the following ATR members for their helpful suggestions and cooperation: Satoshi Koizumi, Toshihiko Shibata, Osamu Sugiyama, Kenta Nohara, Anton Zolotkov. This research was supported by the Ministry of Internal Affairs and Communications of Japan.

References

1. S. Woods et al. Comparing human robot interaction scenarios using live and video based methods, towards a novel methodological approach, *Int. Workshop on Advanced Motion Control*, 2006.
2. A. Green et al., Applying the Wizard-of-Oz framework to cooperative service discovery and configuration, *Proc. IEEE Int. Workshop on Robot and Human Interactive Communication*, 2004.
3. D. A. Norman, *Emotional Design*, Basic Books, New York, 2003.
4. C. T. Ishi, S. Matsuda, T. Kanda, T. Jitsuhiro, H. Ishiguro, S. Nakamura, and N. Hagita, Robust speech recognition system for communication robots in real environments, *IEEE International Conference on Humanoid Robots*, 2006.
5. C. T. Ishi, S. Matsuda, T. Kanda, T. Jitsuhiro, H. Ishiguro, S. Nakamura, et al., Robust speech recognition system for communication robots in real environments. Paper presented at the *IEEE Int. Conf. on Humanoid Robots*, 2006.
6. T. Kanda, H. Ishiguro, M. Imai, and T. Ono, Development and evaluation of interactive humanoid robots, *Proceedings of the IEEE*, Vol. 92, No. 11, pp. 1839–1850, 2004.
7. O. Sugiyama et al., Three-layered draw-attention model for humanoid robots with gestures and verbal cues, *IEEE/RSJ Int. Conf. on Intelligent robots and systems (IROS2005)*, pp. 2140–2145, 2005.
8. T. Kooijmans, T. Kanda, C. Bartneck, H. Ishiguro, and N. Hagita, Accelerating robot development through integral analysis of human-robot interaction, *IEEE Transactions on Robotics*, 2007 (to appear).

6.3 Teleoperation of Multiple Social Robots

Dylan F. Glas, Takayuki Kanda, Member, IEEE,
Hiroshi Ishiguro, Member, IEEE, and Norihiro Hagita, Member, IEEE

ABSTRACT

In this section, we explore the unique challenges posed by the remote operation of multiple conversational robots by a single operator, who must perform auditory multitasking to assist multiple interactions at once. We describe the general system requirements in four areas: social human–robot interaction design, autonomy design, multi-robot coordination, and teleoperation interface design. Based on this design framework, we have developed a system in which a single operator can simultaneously control four robots in conversational interactions with users.

Our implementation includes a graphical interface enabling an operator to control one robot at a time while monitoring several others in the background, and a technique called "Proactive Timing Control," an automated method for smoothly interleaving the demands of multiple robots for the operator's attention. We also present metrics for describing and predicting robot performance, and we show experimental results demonstrating the effectiveness of our system through simulations and a laboratory experiment based on real-world interactions.

6.3.1 Introduction*

In a shared-autonomy approach to robot control, such as that presented in Chapter 6, Section 6.2, improvements in the robot's level of autonomy will reduce the human operator's workload and make it easier to operate the robot. Eventually, if the operator's workload can be sufficiently reduced, it may be possible for that operator to control two or more robots, a situation referred to as single-operator, multiple-robot (SOMR) control.

When semi-autonomous social robots are eventually deployed in real-world situations, SOMR control will be greatly desirable. From a commercial perspective, a fleet of ten service robots controlled by a single human operator would be much more lucrative than the same number of robots requiring a team of twenty operators. The idea of employing a human operator is not unusual—even highly autonomous industrial robots generally have a human in the loop in a supervisory role. Thus, well before full autonomy for social robots is feasible, partial autonomy with a low operator-to-robot ratio could enable social robot applications to be rolled out in the real world.

Our objective is thus to develop techniques enabling SOMR control for partially autonomous social robots, and to steadily decrease the role of the operator over time with improvements in robot technology. As these component technologies improve, the operator-to-robot ratio can be considered as one measure of the degree of autonomy of the system.

In this chapter, we address the unique challenges of SOMR operation for social robots. We examine the key design issues for such systems, and we present an example system we have developed, in which a single operator can monitor and control several conversational robots at once. We also present a mechanism for enabling coordination between different robots, to prevent conflicting demands for an operator's attention.

6.3.1.1 Related Work

In this chapter, we explore semi-autonomous control of multiple robots for social human–robot interaction, by which we mean conversational interaction

---

* This chapter is a modified version of a previously published paper Dylan F. Glas, Takayuki Kanda, Hiroshi Ishiguro, and Norihiro Hagita, Teleoperation of Multiple Social Robots, in *IEEE Transactions on Systems, Man, and Cybernetics—Part A: Systems and Humans/* (accepted for publication, 2011), edited to be comprehensive and fit the context of this book.

between a robot and one or more people. In other fields of robotics, such as search-and-rescue or space exploration, many aspects of both single- and multiple-robot teleoperation are active fields of research, but multiple-robot teleoperation has not yet been studied for social robots.

A substantial amount of work has been done regarding levels of autonomy for teleoperated robots. The concept of "shared autonomy" describes a system in which a robot is controlled by both a human operator and an intelligent autonomous system, a concept that has been used in fields such as space robotics [1] and assistive robotics [2]. The concept of *adjustable autonomy*, also known as *sliding autonomy*, has also been studied, in which varying degrees of autonomy can be used for different situations [3–6].

Other teleoperation research has focused on control interfaces for teleoperation. A wide variety of teleoperation interfaces have been created for vehicle control [8,9], and the unique problems of controlling body position in humanoid robots have also been studied [10].

Several aspects of simultaneous control of multiple robots have also been studied. Hill and Bodt presented field studies observing the effects of controlling multiple robots on operator workload [11]. Sellner et al. studied the situational awareness of an operator observing various construction robots in sequence [12], and Ratwani et al. used eye movement cues to model the situation awareness of an operator supervising several UAVs simultaneously [13].

A key issue in multiple-robot teleoperation is the concept of "fan-out," which describes the number of robots an operator can effectively control [14]. Crandall and Goodrich have laid a theoretical basis for the modeling of SOMR teleoperation, defining metrics such as Interaction Time (IT) and Neglect Tolerance (NT) to help with calculation of robot fan-out and predicting system performance [15]. Thus far, studies of fan-out in multiple-robot teleoperation have focused on tasks such as search and navigation for mobile robots [16], or target selection for UAVs [17], but not social human–robot interaction.

In this chapter, we will build upon this research to define a new application domain: the teleoperation of multiple robots for social human–robot interaction tasks. In doing so, we aim to identify ways in which existing SOMR teleoperation principles can be applied to social robots, and to examine ways in which social robots differ from traditional systems.

6.3.1.2 Design Considerations

In some ways, the teleoperation of multiple robots for social interaction is similar to SOMR teleoperation for conventional robots, and in other ways it presents new challenges. Table 6.2 shows a comparison between social robots and navigational robots with respect to several important concepts for teleoperation. The time criticality of conversational interactions is a central factor affecting many key issues in teleoperation of multiple social robots.

Four key design areas are identified in the system diagram in Figure 6.6. The overall system requirements are driven by the target application, which

TABLE 6.2

Differences in Teleoperation between Navigation (Fundamental Tasks for Mobile Robots [7]) and Social Interaction

	Navigation	Social Interaction (This Study)	New Problems in Social Interaction
Operator's role	Obstacle avoidance. Giving current position, path, goals	Understanding the user's intention and providing required service	
Source of input to operator	Scenery + Map	Audition (+scenery)	Cannot monitor multiple sources simultaneously
Operator's output (low-level control)	Velocity	Utterance, gesture, +(body orientation and position)	Typing and controlling many DOFs for gesturing are very slow
Operator's output (abstracted control)	Position (destination)	Behavior (combination of utterance and gesture)	Difficult to prepare for minor cases in advance
Consequence of ignoring errors caused by autonomy	Crash into obstacle, or lose the robot	Person might get lost, buy wrong product, or receive wrong service	Definitely we should not ignore errors in either case
Can robots wait after an error detected?	Yes, in most cases	No. Users might soon leave if a robot stops	An operator should take control of the robot immediately
Can robots anticipate the timing of possible error?	Not usually	Yes	Most errors are from speech recognition, often after the robot asks a question

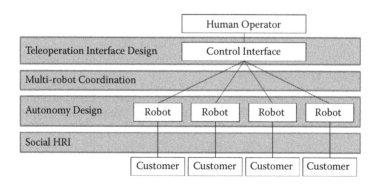

FIGURE 6.6
General overview of multi-robot control system showing key design areas.

in this case falls in the domain of **social human–robot interaction**. This area includes the design of the robot's behavior and dialogue with the goal of creating comfortable, natural, and functional interactions between the robot and a customer (note that we will use the terms *operator* and *customer* here to distinguish between the different human users of the system). To create semi-autonomous robots that can do this, an important issue is **autonomy design**, that is, how operator commands can be reconciled with the autonomous components of the robot control system. Next, due to the time criticality of social interactions, **multi-robot coordination** is necessary to manage the attention of the operator between robots, and to reduce conflicts between demands for the operator's time. Finally, **teleoperation interface design** is necessary to enable interaction between the operator and the robot, providing the operator with situation awareness and controls for operating the robot.

In this section, we will present design considerations in these four areas and propose metrics for quantifying important characteristics of SOMR systems for social interaction.

6.3.1.3 Social Human–Robot Interaction

The target application here is conversational humanlike interaction, primarily focused on dialogue, although nonverbal communication and gestures such as pointing may also be essential interaction components.

Some examples of this type of interaction might include a robot shopkeeper that provides information about various products, an information booth robot that gives directions and answers questions in a shopping mall, a tour-guide robot that explains exhibits in a museum, or a public relations robot that greets people and invites them to visit a shop.

Such simple interactions can be modeled as following a flow that includes alternating phases: one in which a person is asking a question or giving information to the robot, and one in which the robot responds with some explanation or directions. Understanding this pattern of turn-taking defined by the social interaction design helps to enable the coordination of operator attention between multiple robots, as we will explain later.

6.3.1.4 Autonomy Design

In semi-autonomous social robot systems, it is important to define how an operator will interact with the autonomous components of the robot's control system. Generally speaking, an operator can direct high-level tasks or identify errors that the system cannot detect autonomously. For social robots, many necessary functions, such as tracking human positions or presenting information through speech and gesture, can be performed autonomously using available technology. Some core background processes, such as emotional dynamics, can also be automated for social robots [18]. It is in the

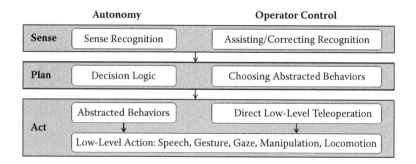

FIGURE 6.7
Autonomy and operator control tasks for sense-plan-act elements in a semi-autonomous robot control system.

recognition and interpretation of verbal and nonverbal communication and the ability to make commonsense judgments based on an understanding of context that an operator can add the greatest value.

For example, an elderly person in a shopping mall who is holding a map and looking around might need route guidance from the robot; on the other hand, a young person in a plaza looking around in a similar manner might just be looking for friends and not need the service. Although a human operator could easily distinguish between these two cases using intuition, visual cues, and implicit social context, such a recognition task would be quite difficult for a robot to perform autonomously.

An operator can provide input to a semi-autonomous system at several levels. Consider a simple framework for robot control, in which developers create sense-plan-act elements based on a pre-assumed world model. Figure 6.7 shows an example of such a system, in which the robot can perform abstracted behaviors composed of low-level actions such as speech and gesture. These behaviors are chosen by decision logic, based on the results of autonomous sensor recognition.

In such a system, three categories of problems tend to occur, which define the three primary tasks of an operator.

6.3.1.4.1 Uncovered Situations

The richness and diversity of human behavior makes it difficult to create a predictive model of the world for social interactions. This can lead to many **uncovered situations**, in which a robot does not have appropriate rules or behaviors implemented to act autonomously. Uncovered situations are of particular concern for systems which interact with humans.

Uncovered situations motivated the original use of the "Wizard-of-Oz" technique, where a dialogue system was controlled by a human operator to collect necessary dialogue elements [19]. By monitoring the interaction, an operator can provide additional information to the system and improve its world model. The assumption behind this technique is that the robot can ultimately cover all situations after collecting a sufficiently complete world model.

6.3.1.4.2 Incomplete Autonomy

Even assuming a good model of the world, there are still cases when we cannot prepare all the necessary sense-plan-act elements. In these cases, an operator can be used as a substitute for **incomplete autonomy** and replace those individual elements. Many WoZ studies in HRI are of this type [20].

An example of replacing a *sense* element is speech recognition. Today's speech recognition technologies are unreliable in noisy environments, as observed by Shiomi et al. in field trials [21]. It is thus not currently possible to automate this sensing task. However, an operator can be employed to listen to the audio stream and manually input the recognized utterances into the system. Using those inputs, the robot can still implement the plan and activate elements autonomously. Other examples in this class could include identifying a person or object, or monitoring the social appropriateness of a robot's actions by observing people's reactions to the robot.

An operator could likewise replace a *plan* element. If a robot's action requires particular expertise or authority, such as that of a doctor, technician, soldier, or law enforcement officer, an operator may be required for this step. Here the robot may be able to sense the environment and act on it, but lack the authority or accountability to make the decision to act.

For replacing an *act* element, an example could be a difficult actuation task such as grasping. The robot might be able to identify an object to grasp and make the decision to grasp it, but need assistance in actually carrying out the grasping task [22].

Note that for this style of teleoperation, the system can often prompt the operator to perform some action. The operator acts as a "black box" in the system, performing some defined processing task on demand, like any other module in the system.

6.3.1.4.3 Unexpected Errors

Finally, it is possible that even if we have prepared a good world model and developed appropriate sense-plan-act elements, the system may not always work as intended. That is, **unexpected errors** may occur during autonomous operation.

In this case, an operator needs to monitor the robot to identify possible errors. In the teleoperation tasks described above, the operator's focus is on the environment and people interacting with the robot, but when monitoring for errors the primary focus is on the performance and behavior of the robot itself.

6.3.1.4.4 Multi-robot Coordination

For conversational interactions, we assume an operator can only correct errors or provide active support for one robot at a time. Particularly in the case of speech recognition, it is extremely difficult for an operator to concentrate on two or more conversations at once. This restriction makes the operator's attention a limited resource.

We thus model a robot's interaction as consisting of **critical sections**, in which there is a high risk of interaction failure and thus a high likelihood that operator assistance will be needed, and **noncritical sections**, which can safely be performed autonomously. Critical sections tend to occur when the actions of the robot depend strongly on recognition of inputs from the customer, and thus the consequences of a recognition error are severe. For example, the moment after a customer asks a question would be a critical section, because the robot needs to respond appropriately to what the customer said. Critical sections can also occur when there is a high probability that an uncovered situation will arise. To prevent interaction failures, it is desirable for an operator to be monitoring a robot during critical sections.

Conflicts arise when two or more robots compete for operator attention by entering critical sections at the same time. Conversational interactions are time critical, and customers may be made to wait if an operator is not present during a critical section. In Section 6.3.2.3, we will propose a mechanism for coordinating the interactions of the robots to eliminate such conflicts.

6.3.1.5 Teleoperation Interface Design

An operator has two tasks to perform: first, supervisory monitoring of all robots to identify unexpected situations, and second, assisting individual robots' recognition, planning, and actuation. Supporting both of these tasks presents a considerable challenge for the user interface design. Both situation awareness and actuation requirements for the user interface differ for these two tasks as follows.

6.3.1.5.1 Monitoring Multiple Robots

When acting in a supervisory role and monitoring multiple robots, the operator needs to identify and react to unexpected problems in a timely manner. A summary of the state information about each robot should be presented to the operator in such a way as to make errors and unusual behavior easily recognizable. Since the highest risk of recognition error occurs during critical sections, alerting the operator to which robots are in or entering critical sections can help manage the operator's attention most effectively.

In some cases, a summary of the robot's state information might not be sufficient for the operator to accurately identify some errors, so it may be important for the operator to periodically examine the detailed state information for individual robots as well.

6.3.1.5.2 Controlling Individual Robots

When controlling a single robot, the operator needs to be aware of the robot's individual situation—with whom the robot is interacting, what that person is saying, and what the robot is doing. For simple systems, such as an information-providing robot in a shopping mall, this immediate information may be sufficient for the robot's interactions. For more elaborate systems where the robot has a long-term relationship with the customer, long-term interaction history or personal information about that customer might be required.

The interface also requires controls for correcting sensor recognition, directing behaviors, and performing low-level control such as entering text for the robot to speak in uncovered situations.

6.3.1.6 Task Difficulty Metrics

Finally, it is valuable to have metrics quantifying the capability of the robot system. For multiple-robot systems, a key quantity is the number of robots a single operator can manage, known as "fan-out." High fan-out can be achieved if the robots can operate with high reliability without the support of an operator, whereas fan-out will be much lower if errors are likely to occur, for example, due to poor sensor recognition or high task difficulty. Thus, to predict fan-out, it is important to have metrics that describe the likelihood of error while the robot is unsupervised. In the terminology of Crandall and Cummings, such metrics are classified as "Neglect Efficiency" metrics [16]. In this section, we will define three neglect efficiency metrics reflecting the risk of interaction errors occurring while the robot is unsupervised. These metrics are summarized in Table 6.3.

6.3.1.6.1 Recognition Accuracy

Sensor recognition accuracy (RA) is a fundamental concern for robots in every field. This is also true for social robots, as recognition of the nuances of communicative signals such as speech, gesture, intonation, and emotion in social interaction can be particularly challenging. An estimate of RA can help predict the frequency of unexpected errors in the "sense" element of the robot's control architecture.

Increasing a robot's RA through better sensors or better recognition technology can reduce the need for operator intervention, and consequently increase the number of robots a single operator can control.

6.3.1.6.2 Situation Coverage

The next metric we propose is Situation Coverage (SC), which describes the completeness of the "plan" and "act" elements in the robot system. We define

TABLE 6.3

Task Difficulty Metrics

Metric	Comments
Recognition Accuracy (RA)	Limited by technology; higher RA increases fan-out
Situation Coverage (SC)	Limited by scenario predictability; higher SC increases fan-out
Critical Time Ratio (CTR)	Determined by interaction design; lower CTR increases fan-out

a situation to be "covered" if the system would autonomously execute the correct behavior given perfect sensor inputs. Using this definition, SC is defined as the percentage of situations encountered by the robot that are covered. SC is useful for describing the upper bound of the system's possible performance, or conversely, a lower bound on the fraction of time during which operator support may be necessary.

In application design, SC is more of a controllable variable than RA. Whereas RA is subject to technological limitations, SC can be increased through human effort. By spending more time researching potential situations the robot may encounter and developing the decision logic and actions to respond to those situations, it is possible to increase a robot's SC.

6.3.1.6.3 Critical Time Ratio

The third metric we will introduce is the Critical Time Ratio (CTR). This is defined as the ratio of the amount of time spent in critical sections to the total duration of an interaction. For tasks with a low CTR, the likelihood of two robots entering a critical section at the same time is correspondingly low, and thus timing control behaviors will seldom be necessary. Tasks with a high CTR are more likely to conflict, which can lead to higher wait times for users and a heavier workload for the operator.

For a designer, it is possible to achieve higher fan-out by creating interactions with a low CTR, for example, by increasing the durations of noncritical sections and minimizing the number of critical sections in the interaction flow. However, this must be done carefully, as reducing CTR also runs the risk of reducing the robot's responsiveness to the customer, and thus reducing the quality of the human–robot interaction.

6.3.2 Implementation

Using these principles, we developed a system for the teleoperation of multiple robots for social interactions. In this section, we will discuss the implementation of our system according to each of the four design areas: social human–robot interaction, autonomy design, multi-robot coordination, and teleoperation interface design.

6.3.2.1 Social Human–Robot Interaction

The interaction flow we developed for this study was based on interactions used in our field trials in a shopping mall. Table 6.4 shows the sequence of conversation phases and their durations. When the robot detected a person in front of it, it would (1) greet the person, and then (2) introduce itself and explain that it can give directions to locations in the shopping mall. After this, the robot would (3) briefly chat about some topic, usually related to current events in the shopping mall or the robot's "experiences" at various shops.

TABLE 6.4

Interaction Sequence

#	Phase	Criticality	Duration
1	Simple greeting	Noncritical	2 s
2	Self-introduction	Noncritical	3 s
3	Chat behavior	Noncritical	Variable
4	Offer guidance	Critical	1s
5	Wait for question	Critical	2–10 s
6	Provide guidance	Noncritical	10–15 s
7	Farewell	Noncritical	5 s

As noted earlier, critical sections include situations where a response from the user is expected, whereas noncritical sections include tasks such as greeting, talking, and giving directions, where the robot is primarily providing information. The critical sections in our flow consist of the robot (4) asking where the customer would like to go (4), and then (5) waiting for the customer's response.

After the question has been asked, the robot (6) gives guidance, and then (7) says goodbye to the customer. All of these phases are considered noncritical.

6.3.2.2 Autonomy Design

We created a semi-autonomous robot control architecture that enables an operator to provide commands and assistance to an otherwise autonomous robot system.

6.3.2.2.1 Robot Control System

The behavior control system used in this study uses a software framework in which short sequences of motions and utterances can be encapsulated into discrete units called "behaviors." The programmer defines a set of transition rules that specify transitions between behaviors based on sensor inputs [23].

With this framework, given perfect Recognition Accuracy and user behavior only within the limits of Situation Coverage, it is possible for the robot to execute any length of behavior chains with full autonomy.

6.3.2.2.2 Operator Intervention

As described in Section 6.3.1.4, the operator needs to be able to intervene in robot operation to deal with uncovered situations, incomplete autonomy, and unexpected errors. This can be achieved either through direct control of the robot at a high or low level, or by correcting the robot's recognition.

Direct control is less desirable, because it bypasses the robot's decision logic, and the operator must take responsibility for both "sense" and "plan" processes. This requires considerable attention and awareness on the part of the operator.

On the other hand, by simply correcting the robot's recognition, the operator only replaces a "sense" element, and the robot handles the "plan" process. The system is less reliant on the operator's awareness and judgment, so this type of intervention is better suited for SOMR control.

6.3.2.3 Multi-Robot Coordination

One fundamental problem with multiple-robot control is that two robots may need an operator's attention at the same time. In such a case, one robot may be making a customer wait for a long time while the operator is helping the other robot. We use two techniques for handling such conflicts.

The first method is to use *conversation fillers* [24]. These are short utterances such as "hmm…" and "please wait a moment," which the robot can automatically say if it is making a customer wait while an operator is not available. This technique can be effective when used in moderation but can leave a negative impression if it is used for too long. For example, when fillers are used after the customer has asked a question, the customer may be frustrated that the robot is not answering.

To avoid this problem, we propose another technique for handling conflicting critical sections, which we call Proactive Timing Control (PTC). This mechanism enables interactions to be coordinated in order to prevent critical section conflicts from arising at all. One means of achieving this is for each robot to send a reservation request to the operator before a critical section begins. If the operator accepts, the robot can proceed to the critical section. Otherwise, the robot performs other behaviors in order to delay entry into the critical section.

With this technique, the delaying behaviors are executed before the user has spoken, while it is still the robot's "turn" in the conversation. The robot has not yet relinquished the initiative, and thus the extra behaviors integrate more smoothly into the flow of interaction. The effectiveness of this technique has been demonstrated in a study of the effects of wait time on customer satisfaction [25].

6.3.2.4 Teleoperation Interface Design

Teleoperation interface software was developed to enable the operator to control one robot (referred to here as the "active" robot) while monitoring the others in the background. The interface used is pictured in Figure 6.8. The four panels on the top left of the screen show the status of each robot, and the operator can click one to begin controlling that robot. The video pane on the upper right shows the video feed from the selected robot's eye camera.

Below those panels, the tabbed button panel on the left can be used to send direct behavior commands to the robot. The column of buttons to the right of that can be used to correct speech recognition results, and the pop-up

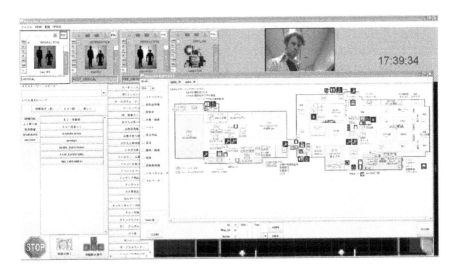

FIGURE 6.8
Teleoperation interface.

window on the right side shows a map of guide destinations from which the operator can trigger guide behaviors.

6.3.3 Experimental Validation

We conducted a laboratory experiment with unbiased participants to evaluate the feasibility and effectiveness of our approach to teleoperation of multiple social robots, as well as the effectiveness of the Proactive Timing Control technique in particular.

6.3.3.1 Laboratory Experiment

6.3.3.1.1 Scenario

The robot's task for this experiment was to provide route guidance to customers. It is easy to imagine a business such as a shopping mall, museum, or theme park placing a robot in a high-visibility location such as a central information booth. This task also lies in an interesting middle-ground between full predictability and open-endedness, and it provides a level of interactivity not found in primarily one-way interactions such as guiding visitors in a museum.

6.3.3.1.2 Experimental Design

The experiment was designed to evaluate the performance of the operator–robot team while varying two factors. The first factor, *robot-number*, was examined at three levels: *2R*, *3R*, and *4R*, representing the number of robots being simultaneously controlled by the operator. We also evaluated two baseline cases: a single-robot case where the operator was always present, referred to as the *1R* condition, and a fully autonomous case with no operator intervention, referred to as the *A* condition. The second factor, *PTC*, was examined at two levels: *with-PTC* and *without-PTC*.

The experiment was designed to evaluate two hypotheses. Our first hypothesis was that our semi-autonomous control scheme would provide better performance than a purely autonomous system, regardless of whether or not PTC was used.

To validate this hypothesis, we tested the performance of our system on an absolute scale, comparing the *A* condition with the others. This comparison was performed separately for the *with-PTC* and *without-PTC* configurations of our system.

Our second hypothesis was that the use of PTC in particular would improve the performance of the robot team, and that this improvement in performance would increase for larger numbers of robots. This was evaluated by a comparison between *with-PTC* and *without-PTC* conditions for each of the *2R*, *3R*, and *4R* cases.

To test these two hypotheses, our experiment included a total of 8 conditions to be evaluated: *with-PTC* and *without-PTC* variations for each of the *2R*, *3R*, and *4R* cases, and the two baseline cases, for which the use of PTC is not relevant.

6.3.3.1.3 Setup

The behaviors and decision logic for the route guidance scenario were adapted from a deployment of our robots in a shopping mall. We used the interaction flow described in Section 6.3.2.1, with the chat behaviors (Phase 3 in Table 6.4) adapted for use as PTC behaviors.

The PTC behaviors consisted of interruptible sequences of short behaviors with an average duration of 4.4 s. After each behavior, the sequence could be interrupted or continued based on the presence or absence of an operator.

An example of such a sequence is the following: "Hi, I'm Robovie. / I know many things about this shopping mall. / This week the mall is having a special anniversary celebration. / There are many discount campaigns and exciting activities planned! / There is a 10% off sale in the clothing section. / And next Sunday there will be a classical music concert!" When an operator became available, the sequence could be interrupted after any of these utterances so the robot could begin the critical section by offering to give route guidance, and the entire sequence would flow in a fairly natural way. Four of these PTC sequences were prepared for the experiment, with a maximum possible length of 12 behaviors each, and one sequence was chosen at random for each interaction.

It should be noted that these behaviors were not irrelevant time-killing behaviors. These chat behaviors had originally been part of the natural conversation flow. When the robot spoke about these topics in the field trial, they were relevant to the customers, who enjoyed their interactions with the robot.

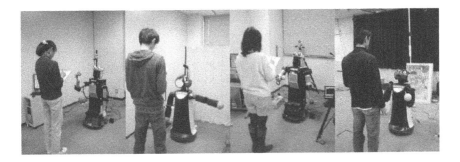

FIGURE 6.9
Four robots operated simultaneously in our experiment.

6.3.3.1.4 Robots

The experiment was conducted in our laboratory, using two Robovie-II and two Robovie-R2 robots, as shown in Figure 6.9. Each robot was set up with an ATRASR automatic speech recognition (ASR) system [26], which operated in parallel with the operator.

6.3.3.1.5 Participants

Sixteen paid participants played the role of customers in this experiment (12 male, 4 female, average age 22.3, SD = 2.5 years). All were native Japanese speakers.

One expert operator, an assistant in our laboratory, was employed to control the robots for all trials. The operator was trained in the use of the control interface and thoroughly familiar with the map of guide destinations prior to the experiment, so we assume negligible improvement in operator performance across trials.

6.3.3.1.6 Procedure

To provide the operator with consistent task difficulty in the different experimental conditions, each trial consisted of 24 interactions in total, that is, 6 interactions per robot in the *4R* case, 8 in the *3R* case, 12 in the *2R* case, and 24 in the *1R* case. The *A* condition was conducted with four robots but no operator.

Each interaction included a greeting from the robot, possible PTC chat behaviors, a question from the customer, and a response and farewell from the robot. Eight trials were run on each day of the experiment, one for each condition (*2R-with, 2R-without, 3R-with, 3R-without, 4R-with, 4R-without, 1R,* and *A*).

On the customer side, four participants took part in every trial, and each participant interacted with the robots a minimum of 6 times per trial. Participants were assigned evenly across the robots. To achieve even distribution in the *3R* conditions, three participants interacted 6 times each with assigned robots, while one participant moved between the robots, performing two interactions with each. In other conditions, participants did not move between robots.

This experimental procedure was repeated on four days with a different group of 4 customer participants on each day, for a total of 16 participants acting as customers. The order of the eight trials on each day was counterbalanced with respect to both the *robot number* and *PTC* factors.

For consistency in timing, interactions were robot-initiated, with the robot inserting a pause of 0–5 s between interactions. To provide a consistent level of workload for the operator, participants continued interacting with the robots for the entire duration of each trial, going beyond the 6 evaluated interactions if necessary.

■ With PTC   □ Without PTC

6.3.3.1.7 Evaluation

There is a chain of effects which we expect to produce different results between the *with-PTC* and *without-PTC* conditions. First, the use of PTC should increase the number of critical sections for which the operator is present. This should consequently increase the interaction success rate, because the speech recognition system is used less often. Finally, this improved success rate combined with reduced wait time in the critical section should improve customer satisfaction.

Accordingly, to evaluate the performance of the system, we measured three variables: the rate of operator supervision in the critical section, the overall ratio of successful interactions, and customer satisfaction on a scale of 1 (unsatisfied) to 7 (satisfied). Interaction success (whether the robot had successfully answered the question) and customer satisfaction were reported by participants after each interaction.

6.3.3.2 Experimental Results

The results of this experiment are illustrated in Figure 6.10a, showing operator supervision during the critical sections; Figure 6.10b, showing the interaction success rates; and Figure 6.10c, showing results from the customer satisfaction questionnaire.

6.3.3.2.1 Absolute Comparison

To evaluate the absolute performance of the system between *with-PTC* and *without-PTC*, we examined each *PTC* condition separately, comparing the *2R*, *3R*, and *4R* levels of that condition with the *1R* and *A* baseline cases.

Operator supervision in critical section: Due to the use of PTC, the operator availability during critical sections was 100% for every trial in the *with-PTC* condition (Figure 6.10a). In the *without-PTC* condition, operator availability decreased markedly as the number of robots increased.

Interaction success: For the *with-PTC* conditions, 100% of the robot's responses were correct, which is to be expected as the operator was present for all interactions. In the *without-PTC* conditions, the interaction success

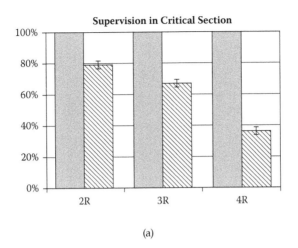

(a)

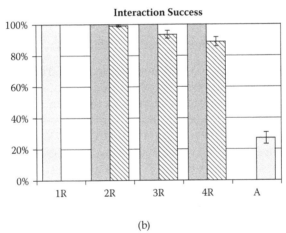

(b)

FIGURE 6.10

Results from our experimental evaluation. (a) Operator supervision for critical sections. (b) Interaction success rate. (c) Customer satisfaction. Error bars show standard error.

rate decreased as the number of robots increased, up to a 10% failure rate in the *4R* condition.

In both conditions, there was a significant difference when compared with the autonomous case, which was successful only 27% of the time (*with-PTC* condition: $\chi^2(4) = 327.805$, $p < .01$, residual analysis: *1R, 2R, 3R*, and *4R* to *A*: $p < .01$, *without-PTC* condition: $\chi^2(4) = 247.307$, $p < .01$, residual analysis: *1R, 2R*, and *3R*, to *A*: $p < .01$, and *4R* to *A*: $p < .05$).

Customer satisfaction: For the *with-PTC* condition, customer satisfaction did not vary significantly between the *1R – 4R* conditions. A repeated-measures ANOVA revealed a significant difference in the main effect of *robot number* ($F(4,15) = 189.786$, $p < .001$). A Bonferroni test revealed *1R, 2R, 3R*, and *4R* to be

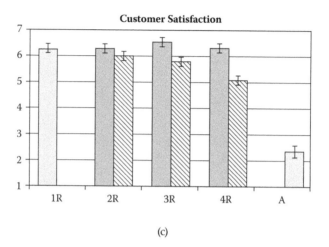

(c)

FIGURE 6.10 (continued)

significantly better than A ($p < .001$), but no significant difference was found among *1R, 2R, 3R,* and *4R.*

For the *without-PTC* condition, customer satisfaction did not vary significantly between the *1R – 3R* conditions, but decreased at *4R.* A repeated-measures ANOVA revealed a significant difference in the main effect of number of robots ($F(4,15) = 108.571, p < .001$). A Bonferroni test revealed that *1R, 2R, 3R,* and *4R* were significantly better than *A* ($p < .001$), and *1R* and *2R* were significantly better than *4R* ($p < .001$ and $p < .01$). The difference between *3R* and *4R* was approaching significance ($p = .077$). There were no significant differences among *1R, 2R,* and *3R.*

These results confirm our hypothesis that performance in all teleoperated cases would be higher than the autonomous baseline. For the *4R* case, the significant decrease in customer satisfaction for the *without-PTC* condition also agrees with our prediction.

6.3.3.2.2 Relative Comparison

To confirm to the relative effect of PTC, we directly compared the customer satisfaction for *with-PTC* and *without-PTC* for each level of the number of robots. A paired *t*-test revealed significant differences for *3R* ($t = 4.442, p < .001$), and *4R* ($t = 4.986, p < .001$), and an almost significant difference for *2R* ($t = 1.813, p = .090$).

This result is consistent with our hypotheses that the use of PTC will improve performance, and that the performance improvement will be stronger for larger numbers of robots.

6.3.3.3 Operator Experience

During this experiment, the operator often remarked that she felt a high level of pressure and frustration during the trials without PTC, because she was aware that many robots were entering critical sections at the same time. She said she felt relaxed, and that the interactions seemed to go smoother when PTC was used.

6.3.4 Simulation

Our laboratory trials provided a practical demonstration of a single operator controlling multiple robots in conversational interactions, but such trials are expensive in terms of time and resources, and it is logistically difficult to evaluate performance with large numbers of robots. In order to further explore the dynamics of PTC and to make projections about the performance of our system under a variety of conditions, we created a simulation based on the interactions observed in our experiment.

6.3.4.1 Interaction Model

The interaction model used in the simulation represents each interaction as a sequence of phases, as shown in Table 6.5. Interactions normally proceed in sequence through the Pre-Critical, Critical Section, Post-Critical, and Non-Interacting phases. If Proactive Timing Control is being used, then the system will transition to a PTC Behavior rather than a Critical Section if the operator is unavailable. The length of each phase is modeled as a normal distribution with mean and standard deviation calculated from the interactions conducted in our experiment.

6.3.4.2 Task Success

Task success is estimated by categorizing each Critical Section as attended or unattended. For our simulation, if an operator is present for an entire Critical Section, it is considered to be attended. If the operator is absent for any fraction of the Critical Section, it is considered to be unattended. This method of counting is used because it is important to attend a Critical Section from the beginning in order to guarantee that the customer's question is heard in its entirety. If the operator is late, the speech recognition system may have

TABLE 6.5

Interaction Phases and Durations

Interaction Phase	Mean Duration(s)	Standard Deviation(s)
Precritical	4.9	1.1
PTC Behavior	4.4	1.7
Critical Section	6.3	5.0
Postcritical	14.8	2.8
Noninteracting	0.5	0.0

already provided an incorrect response, or the operator may need to repeat the question.

In our experiment, the operator's accuracy rate during attended interactions was 100%, whereas the ASR system's accuracy in the autonomous case was 27%. Our simulation thus assumes a response accuracy of 100% for attended interactions and 27% for unattended interactions.

6.3.4.3 Operator Allocation

The simulated operator is allocated to robots according to the following simple algorithm:

If the operator's current robot is in a critical section, do not switch to a new robot.

Otherwise, if any other robot is currently in a critical section, switch to the robot which has been in its critical section the longest.

Otherwise, switch to the robot for which the anticipated critical section begins soonest.

This algorithm is not necessarily guaranteed to be optimal, but it is roughly based on the way operators were observed to operate the system during testing.

Figures 6.10 and 6.11 show example simulated interactions that illustrate how PTC dramatically reduces the number of unattended critical sections. The simulated operator in Figure 6.11 is only present for the beginning of 31% of critical sections, whereas the operator in Figure 6.12 is present for 100%.

6.3.4.4 Number of PTC Behaviors

As the number of robots increases, more PTC behaviors will be required, and the average length of interactions will increase. We examined this trend using our simulation.

Figure 6.13 shows the maximum and average number of PTC behaviors used by our simulated system in runs of 1000 interactions using 1–8 robots. Here, one PTC behavior consists of a short utterance of around 4.4 s in length.

The results from this simulation agreed closely with our experimental results, as our operator used a maximum of 10 and an average of 3.9 PTC behaviors for the 4-robot case, compared with a maximum of 10 and average of 3.1 in the simulation.

As discussed in Section 6.3.1.2, Critical Time Ratio (CTR) is determined by the design of an interaction. A highly interactive robot application would have long critical sections and thus a high CTR, whereas a robot mostly performing fixed behaviors with less responsiveness to a customer would have a low CTR. Figure 6.14 shows the average number of PTC behaviors used in our simulations for interactions using a base CTR (not including PTC

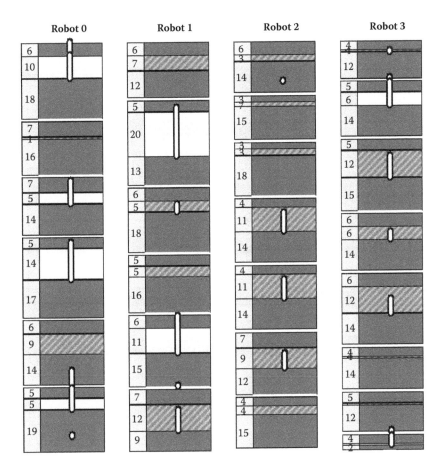

FIGURE 6.11
Examples of simulated interactions *without* Proactive Timing Control. Dark gray boxes represent noncritical interaction phases. Light-colored boxes represent attended Critical Sections, and diagonally hatched red boxes represent unattended Critical Sections. Numbers to the left of each phase indicate its duration in seconds. Vertical bars indicate which robot the operator is attending to at any given time.

behaviors) ranging from 0.1 to 0.5. The figure illustrates how an interaction designer can balance the CTR of an interaction with the desired average PTC duration to target a given number of robots.

6.3.4.5 Relying on Autonomy

The results so far assume an unlimited number of PTC behaviors and a target of perfect operator attendance during critical sections. However, the choice of how many PTC behaviors to use can be seen as a trade-off between the desired level of response accuracy and its cost in terms of design difficulty and extended interaction time. Limiting the number of PTC behaviors

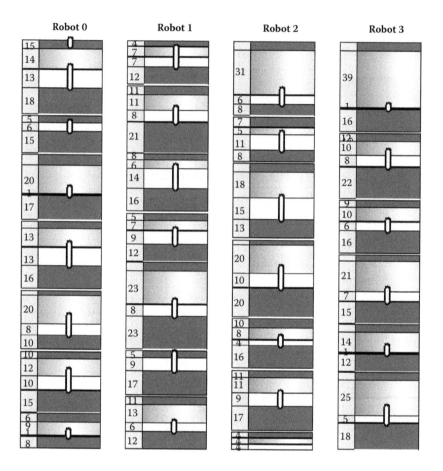

FIGURE 6.12

Examples of simulated interactions *with* Proactive Timing Control. Dark gray boxes represent noncritical interaction phases. Light-colored boxes represent attended Critical Sections, and boxes with metallic shading represent PTC delay behaviors. Numbers to the left of each phase indicate its duration in seconds. Vertical bars indicate which robot the operator is attending to at any given time.

causes the system to rely more on autonomy, which increases the risk of error. However, as the capabilities of recognition systems improve over time, it may be possible to rely more heavily on autonomy and thus achieve very high performance with minimal use of PTC.

6.3.5 Discussion

We were quite pleased by the positive results of this study. The results from our experiment showed our approach to multi-robot control for conversational interactions to be quite effective, and we have subsequently applied these techniques in a number of field trials and laboratory studies.

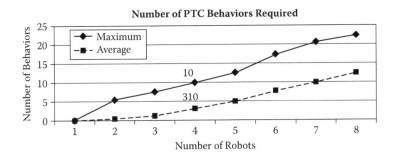

FIGURE 6.13
As the number of robots increases, more PTC behaviors are required to guarantee that an operator can attend all Critical Sections.

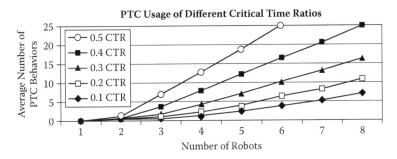

FIGURE 6.14
The average number of PTC behaviors required for a given number of robots increases as a function of Critical Time Ratio.

6.3.5.1 Effectiveness of Shared Autonomy

As the simulation results illustrate, the maximum number of robots an operator can control depends on a variety of factors, including sensor reliability, critical time ratio, maximum number of PTC behaviors, and acceptable error rate. For the most difficult interaction settings in our experiment, the operator was successfully able to control four robots with 90% task success, and for the trials using PTC the operator was 100% successful in conducting all 288 interactions with no errors. Both of these results are dramatically superior to the low 27% success rate of the robots operating autonomously.

6.3.5.2 Defining Criticality

One conceptual model contributing to the success of our system was the division of interactions into critical and noncritical sections. This model can be applied to many kinds of interactions, such as providing information, giving directions, and providing services requested by a customer. It can also be adapted for more complex interactions. For example, if a robot needs to ask a series of several questions, it may make sense to extend the critical section to encompass all of them in a single block. This may result in a small amount of wasted time for the operator while the robot is giving explanations or asking questions, but the operator is also guaranteed to be present for each of the follow-up questions, at a time where it may be awkward to insert delay behaviors.

In the general case, it may be useful to consider both the risk of error and the cost of that error, both of which can be continuous variables. These subtleties may become more important in complex or long-term interactions; however, for the simple interactions in this study, we will consider only two levels of criticality and model all failures as having equal cost.

6.3.5.3 Operator Workload

This study has focused on using an operator at maximum efficiency to control several robots, but little attention has been paid to the operator's workload, stress level, or fatigue. In the *4R* conditions of our laboratory experiment, the operator was almost continuously busy with time-critical tasks, with no opportunity to rest. This kind of operation cannot be sustained for long periods of time.

Within these high-workload conditions, the use of PTC made a big difference to the operator's experience. In the *with-PTC* condition, the operator's tasks were assigned sequentially: a task never started until the operator had finished the previous task. In the *without-PTC* condition, task requests were often sent to the operator while she was busy helping a different robot, and she reported that she felt these situations to be stressful and frustrating.

6.3.5.4 Limitations

Our experiment was conducted in a laboratory with 16 participants repeatedly interacting with the robots. The use of PTC in real-world deployments of robots may have a stronger or weaker effect on customer satisfaction due to factors such as the novelty effect of the robots, customers' lack of familiarity with the robot's conversation style due to nonrepeated interactions, quality and appropriateness of the robot's utterances, and variation based on the deployment context, for example, whether people in that environment are in a relaxed or rushed mood.

6.3.6 Conclusions

In this study, we have presented a general framework for enabling the simultaneous teleoperation of multiple social robots, focusing on four key design areas: human–robot interaction design, autonomy design, multiple-robot coordination, and teleoperation interface design. While many key aspects of autonomy design and teleoperation interface design are similar to issues faced in other fields of robotics, the areas of human–robot interaction design and multi-robot coordination present many new issues that are unique to social robots.

Based on this conceptual framework, we implemented a robot system to demonstrate the new concept of a single operator controlling multiple robots in simultaneous social interactions. Our laboratory evaluations showed our system to be quite successful, with an operator achieving over 95% task success while controlling up to four robots in one experiment. These results demonstrate the value of our conceptual framework as well as the effectiveness of our specific solutions, such as Proactive Timing Control.

In our experiment, task success and customer satisfaction in every condition were far superior to those attainable by the same system operating in a fully autonomous mode. Furthermore, our simulation results show that PTC reduces or eliminates conflicts between robots for an operator's attention. Even when PTC behaviors are limited and the operator is forced to rely on automatic speech recognition some of the time, our simulation results indicate that PTC will provide a substantial increase in task success over a system with no timing control.

Most important, we have tested this system using an actual task often performed by our robots in the field, suggesting that this technology can be immediately put to use in real-world field trials. This study introduces the new field of teleoperation for multiple social robots, and several of the topics addressed in this chapter are promising areas for further in-depth research.

Acknowledgment

This work was supported by the Ministry of Internal Affairs and Communications of Japan.

References

1. B. Brunner, G. Hirzinger, K. Landzettel, and J. Heindl, Multisensory shared autonomy and tele-sensor-programming—key issues in the space robot technology experiment ROTEX, *IEEE/RSJ Int. Conf. on Intelligent Robots and Systems (IROS)*, 1993, pp. 2123–2139.

2. D. Vanhooydonck, E. Demeester, M. Nuttin, and H. Van Brussel, Shared control for intelligent wheelchairs: an implicit estimation of the user intention, *Proc. of the 1st International Workshop on Advances in Service Robotics (ASER)*, Bardolino, Italy, March 2003, pp. 176–182.

3. D. B. Kaber and M. R. Endsley, The effects of level of automation and adaptive automation on human performance, situation awareness and workload in a dynamic control task, *Theoretical Issues in Ergonomics Science*, Vol. 5, No. 2, March–April 2004, pp. 113–153.

4. M. A. Goodrich, T. W. McLain, J. D. Anderson, J. Sun, and J. W. Crandall, Managing autonomy in robot teams: observations from four experiments. *ACM/IEEE 2nd Annual Conference on Human-Robot Interaction (HRI2007)*, 2007, pp. 25–32.

5. P. Scerri, D. Pynadath, and M. Tambe, Towards adjustable autonomy for the real-world, *Journal of AI Research (JAIR)*, 2002, Vol. 17, pp. 171–228.

6. B. P. Sellner, F. Heger, L. Hiatt, R. Simmons, and S. Singh, Coordinated multi-agent teams and sliding autonomy for large-scale assembly, *Proc. of the IEEE—Special Issue on Multi-Robot Systems*, Vol. 94, No. 7, July 2006, pp. 1425–1444.

7. A. Steinfeld, T. Fong, D. Kaber, M. Lewis, J. Scholtz, A. Schultz, M. Goodrich, Common metrics for human-robot interaction, *ACM/IEEE 1st Annual Conference on Human-Robot Interaction (HRI2006)*, pp. 33–40, 2006.

8. T. Fong and C. Thorpe, Vehicle teleoperation interfaces, *Autonomous Robots*, Vol. 11, No. 1, 2001, pp. 9–18.

9. H. H. Chiang, S. J. Wu, J. W. Perng, B. F. Fu, and T. T. Lee, The human-in-the-loop design approach to the longitudinal automation system for an intelligent vehicle, *IEEE Transactions on Systems, Man, and Cybernetics Part A: Systems and Humans*, Vol. 40, No. 4, July 2010, pp. 708–720.

10. N. E. Sian, K. Yokoi, S. Kajita, and K. Tanie, Whole body teleoperation of a humanoid robot integrating operator's intention and robot's autonomy, *IEEE/RSJ Int. Conf. on Intelligent Robots and Systems (IROS2003)*, 2003, pp. 1651–1656.

11. S. G. Hill and B. Bodt, A field experiment of autonomous mobility: operator workload for one and two robots, *ACM/IEEE 2nd Annual Conference on Human-Robot Interaction (HRI2007)*, 2007, pp. 169–176.

12. B. P. Sellner, L. M. Hiatt, R. Simmons, and S. Singh, Attaining situational awareness for sliding autonomy, *ACM/IEEE 1st Annual Conference on Human-Robot Interaction (HRI2006)*, 2006, pp. 80–87.

13. R. M. Ratwani, J. M. McCurry, and J. G. Trafton, Single operator, multiple robots: an eye movement based theoretic model of operator situation awareness, *Proc. 5th ACM/IEEE International Conference on Human-Robot Interaction (HRI2010)*, 2010, pp. 235–242.

14. D. R. Olsen, Jr. and S. B. Wood, Fan-out: measuring human control of multiple robots, *Proc.of the SIGCHI conference on Human factors in computing systems*, Vienna, Austria, April 24–29, 2004, pp. 231–238.

15. J. W. Crandall and M. A. Goodrich, Characterizing efficiency of human robot interaction: a case study of shared-control teleoperation, *IEEE/RSJ International Conference on Intelligent Robots and Systems (IROS2002)*, 2002, pp. 1290–1295.

16. J. W. Crandall, M. L. Cummings, Developing performance metrics for the supervisory control of multiple robots, *ACM/IEEE 2nd Annual Conference on Human-Robot Interaction (HRI2007)*, 2007, pp. 33–40.

17. M. L. Cummings and P. J. Mitchell, Predicting controller capacity in remote supervision of multiple unmanned vehicles, *IEEE Systems, Man, and Cybernetics, Part A Systems and Humans*, Vol. 38, No. 2, 2008, pp. 451–460.

18. M. Álvarez, R. Galán, F. Matía, D. Rodríguez-Losada, and A. Jiménez, An emotional model for a guide robot, *IEEE Transactions on Systems, Man, And Cybernetics Part A: Systems and Humans*, Vol. 40, No. 5, September 2010, pp. 982–992.

19. N. Dahlbäck, A. Jönsson and L. Ahrenberg, Wizard of Oz studies: why and how, *International Conference on Intelligent User Interfaces*, 1993, pp. 193–200.

20. A. Steinfeld, O. C. Jenkins, and B. Scassellati, The Oz of Wizard: simulating the human for interaction research, *4th ACM/IEEE International Conference on Human-Robot Interaction (HRI2009)*, 2009, pp. 101–107.

21. M. Shiomi, D. Sakamoto, T. Kanda, C. T. Ishi, H. Ishiguro, and N. Hagita, A semi-autonomous communication robot—a field trial at a train station, *ACM/IEEE International Conference on Human-Robot Interaction (HRI2008)*, 2008, pp. 303–310.

22. N. Sian, T. Sakaguchi, K. Yokoi, Y. Kawai, and K. Maruyama, Operating humanoid robots in human environments, *Proc. RSS Workshop: Manipulation for Human Environments*, Philadelphia, PA, Aug. 2006.

23. T. Kanda, H. Ishiguro, M. Imai, and T. Ono, Development and evaluation of interactive humanoid robots, *Proceedings of the IEEE*, Vol. 92, No. 11, pp. 1839–1850, 2004.

24. T. Shiwa, T. Kanda, M. Imai, H. Ishiguro, and N. Hagita, How quickly should communication robots respond?, *ACM/IEEE 3rd Annual Conference on Human-Robot Interaction (HRI2008)*, Amsterdam, the Netherlands, 2008, pp. 153–160.

25. K. Zheng, D. F. Glas, T. Kanda, H. Ishiguro, and N. Hagita, How many social robots can one operator control?, *ACM/IEEE 6th Annual Conference on Human-Robot Interaction (HRI 2011)*, Lausanne, Switzerland, 2011, pp. 379–386.

26. T. Shimizu et al., Spontaneous dialogue speech recognition using cross-word context constrained word graph, *Proc. ICASSP*, 1996, pp.145–148.

7

Learning and Adaptation

7.1 Introduction

Social robots work closely with people in daily life. Different people have different preferences, and robots need to work with all of them. Moreover, there are diverse environments robots would need to work with. Thus, a robot will need to adapt itself to people and environments when deployed. Perhaps some of this adaptation could be done by a human engineer or operator in advance; though, if it requires careful observation, real-time adjustment, and working with a lot of data, it is not realistic to rely on humans. It seems clear that social robots need to be equipped with learning capabilities to adapt themselves to people and environments.

In contrast, previous research in robotics had typically focused on learning with nonhuman entities, such as learning maps [1] and meanings of places [2–4], and learning task-specific motions by imitation [5,6]. The obvious difference with such previous techniques is the fact that humans are entities who have complex internal states that are difficult to observe, and their behaviors are too complex to predict. Thus, learning and adaptation in human–robot interaction is a highly challenging problem.

For this role, we have to densely distribute sensors and precisely monitor both robots and humans. Early works primary analyzed how human's preferences can be inferred from their behavior. For instance, proximity was often studied. It was revealed that people keep similar distances from robots that they do from other people [7,8]; thus, as found in a human communication study [9], we could infer psychological proximity from physical proximity. Moreover, it is revealed that people's other movements, such as cooperative movements, are correlated with their subjective preference for a robot [10]. This might also echo the findings in human communication that people show similar movements as a partner if they want to be friendly with him or her [11].

There are techniques to adapt robots' behavior. For instance, these works started to reveal ways to learn preferences regarding proximity [12], behavior pattern [13], and walking speed [14]. Some studies rely on signals directly observed by the robot, whereas others rely on signals received from external devices [12,13].

This chapter introduces three cases of learning and adaptation in human–robot interaction. In the first study, we introduce a work on improving time-critical adaptation. As introduced in Chapter 3, some people are nervous about interacting with a social robot. There are a number of potential reasons, but such negative attitudes can be observed from people's behavior toward a robot [15]. Once a robot observes such nervousness, what should it do? The second study shows a way to moderate people's tension in interacting with a robot.

In the second study, we introduce a work related to adaptation of robot's behavior parameters to individuals' preferences. We expect that people have different preference regarding movements and behavior, such as interaction distances, eye contact, and the speed and timing of movements. To make a robot adaptive to such preferences, we employed a learning method in combination with a method to infer people's subjective preference toward a robot [10].

The third study is about adaptation toward people in the environment. When we put a robot in a public space, it would need to adapt itself to the space. Some basic information to know, for example, is maps and places of interest. Moreover, we need to know people in the environment. For example, if a robot intends to talk to a person, it needs to know where people walk. It involves understanding the meaning of space as well as people's global behavior in the environment. The study reveals a way to retrieve such information by observing people's movements in the environment.

Three studies deal with adaptations at different levels. While the first study addresses events occurring within a few minutes of interaction, the second study potentially deals with long-term adaptations, in which a single individual interacts repeatedly over time. The third study addresses adaptation toward rather macroscopic phenomena, that is, adaptation to a set of people who visit the same location. Such adaptation to a set of people is essentially enabled by the acquirement of common understanding of the environment implicitly shared among people.

Reference

1. Fox, D., Burgard, W., and Thrun, S. (1999). Markov localization for mobile robots in dynamic environments. *Journal of Artificial Intelligence Research, 11*, 391–427.
2. Mozos, Ó. M., Triebel, R., Jensfelt, P., Rottmann, A., and Burgard, W. (2007). Supervised semantic labeling of places using information extracted from sensor data. *Robotics and Autonomous Systems, 55*(5), 391–402.
3. Spexard, T., Li, S., Wrede, B., Fritsch, J., Sagerer, G., Booij, O. et al. (2006). BIRON, where are you? Enabling a robot to learn new places in a real home environment by integrating spoken dialog and visual localization. Paper presented at the *IEEE/RSJ Int. Conf. on Intelligent Robots and Systems (IROS2006)*.

4. Vasudevan, S., Gachter, S., Nguyen, V., and Siegwart, R. (2007). Cognitive maps for mobile robots–an object based approach. *Robotics and Autonomous Systems, 55*(5), 359–371.

5. Atkeson, C. G., and Schaal, S. (1997). Robot learning from demonstration. Paper presented at the *4th International Conference on Machine Learning*.

6. Billard, A., and Dautenhahn, K. (1999). Experiments in learning by imitation—grounding and use of communication in robotic agents. *Adaptive Behavior, 7*(3), 411–434.

7. Dautenhahn, K., Walters, M. L., Woods, S., Koay, K. L., Nehaniv, C. L., Sisbot, E. A. et al. (2006). How may I serve you? A robot companion approaching a seated person in a helping context. Paper presented at the *ACM/IEEE Int. Conf. on Human-Robot Interaction (HRI2006)*.

8. Hüttenrauch, H., Eklundh, K. S., Green, A., and Topp, E. A. (2006). Investigating spatial relationships in human-robot interactions. Paper presented at the *IEEE/RSJ Int. Conf. on Intelligent Robots and Systems (IROS2006)*.

9. Hall, E. T. (1966). *The Hidden Dimension: Man's Use of Space in Public and Private.* The Bodley Head Ltd.

10. Kanda, T., Ishiguro, H., Imai, M., and Ono, T. (2003). Body movement analysis of human-robot interaction. Paper presented at the *International Joint Conf. on Artificial Intelligence (IJCAI2003)*.

11. Chartrand, T. L., and Bargh, J. A. (1999). The Chameleon effect: The perception-behavior link and social interaction. *Journal of Personality and Social Psychology, 76*(6), 893–910.

12. Walters, M. L., Oskoei, M. A., Syrdal, D. S., and Dautenhahn, K. (2011). A long-term human-robot proxemic study. Paper presented at the *IEEE Int. Symposium on Robot and Human Interactive Communication (RO-MAN2011)*.

13. Suga, Y., Endo, C., Kobayashi, D., Matsumoto, T., Sugano, S., and Ogata, T. (2006). Adaptive human-robot interaction system using interactive EC. Paper presented at the *IEEE/RSJ Int. Conf. on Intelligent Robots and Systems (IROS2006)*.

14. Sviestins, E., Mitsunaga, N., and Kanda, T. (2007). Speed adaptation for a robot walking with a human. Paper presented at the *ACM/IEEE Int. Conf. on Human-Robot Interaction (HRI2007)*.

15. Nomura, T., Kanda, T., Suzuki, T., and Kato, K. (2008). Prediction of human behavior in human-robot interaction using psychological scales for anxiety and negative attitudes toward robots. *IEEE Transactions on Robotics, 24*(2), 442–451.

7.2 Moderating Users' Tension to Enable Them to Exhibit Other Emotions

Takayuki Kanda,[1] Kayoko Iwase,[1] Masahiro Shiomi,[1,2] and Hiroshi Ishiguro[1,2]

[1]*Department of Communication Robot, ATR Intelligent Robotics and Communication Laboratories,* [2]*Faculty of Engineering, Osaka University*

ABSTRACT

We propose a system to adapt a robot's interactive behavior to moderate users' tension, and enable users to exhibit emotions in interaction. Previous studies in human–robot interaction revealed that users tend to exhibit negative emotions, particularly tension, when they interact with a robot. While users' emotion could be used in recognition to sense their attitude during interaction, such tension would prevent users from exhibiting emotions during interaction. We analyzed whether tension is exhibited during interaction. The analysis reveals that users do exhibit tension, and it prevents users from exhibiting other automatic emotions caused by a stimulus, such as joy and fear. Based on these findings, we developed a robot system to adapt to users' tension. When tension is exhibited, the robot exhibits tension-moderating behavior to alleviate users' tension. When no such tension is observed, the robot incorporates the result from emotion recognition to supplement speech recognition, which is not robust enough by itself due to noise in perceived speech from a distance.

7.2.1 Introduction*

While we discussed research works on behavioral naturalness in the Chapter 4, we observed that a large amount of information is exchanged nonverbally in human communication. For instance, Mehrabian revealed that 93% of the impression from a message (e.g., positive or negative) is conveyed nonverbally, while 7% of it is conveyed verbally [1]. How can such subtle information exchange be addressed in human–robot interaction?

Expression of emotions has been well studied [2,3]. Further, there are robots that infer users' state from emotion expressed by users, such as for creating affective reaction [4], and estimating of user context [5]. Fujie et al. utilized paralinguistic information and motion, such as nodding, to recognize the attitude of a user from the message expressed. Komatani et al. implemented an emotion recognition function into a humanoid robot and found that people often felt tense [6].

As a study shown in Chapter 3.3 reported, some users have a negative attitude, for example, hesitation, toward interacting with a robot. It seems that tension emotion is one of the signs showing such a negative attitude. We consider that working with tension emotion will be a useful first step toward adapting robots to individuals. In this chapter, we propose a method to detect and moderate tension to promote interaction with a robot. It also utilizes other detected emotion to supplement the speech recognition ability

---

* This chapter is a version of a previously published paper (Kanda, T., Iwase, K., Shiomi, M. and Ishiguro, H., 2005, A tension-moderating mechanism for promoting speech-based human–robot interaction, IEEE/RSJ International Conference on Intelligent Robots and Systems (IROS2005), pp. 527–532.), modified to be comprehensive and fit the context of this book.

of the robot, which is deficient because people typically speak at 50–100 cm distant from a robot, which causes considerable noise in the speech signal.

7.2.2 Does Tension Disturb Occurrence of Other Emotions?

7.2.2.1 Background

There are many research works on emotions. Ekman argued the existence of basic emotions, which are common emotions in humans, and proposed six basic emotions [7]: anger, disgust, fear, joy, sadness, and surprise. Russell assumed the two basic dimensions of emotions, "pleasure–unpleasure" and "low arousal–high arousal," and proposed circumplex model (Figure 7.1), where the six basic emotions and other emotions are mapped on a circle [8].

Emotions have been classified relative to time [9]. The shortest one is autonomic emotion mostly caused by a stimulus (e.g., an utterance from a robot) and lasts seconds, such as joy and fear. Another is self-reported emotion, which humans are conscious of and can describe. Self-reported emotion is relatively independent, with such a single stimulus and lasts minutes, such as tension. (Others are related to moods and emotional disorders changes in hours, days, and so forth.)

7.2.2.2 Hypothesis: Disturbance Caused by Tension

Russell's circular model implies that if there is a strong emotion expressed, the other emotions may not be observed (Figure 7.1). As Komatani et al. reported, people often felt tense during human–robot interaction [6]. This connects with the study conducted by Nomura et al. that reported that people who have a negative attitude toward a robot tend to avoid communication with it [10].

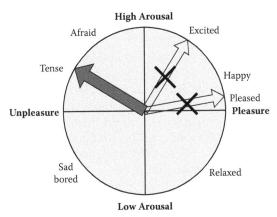

FIGURE 7.1
Hypothesis on the disturbance by tension emotion based on the Circumplex model proposed by Russell.

Based on these findings, we consider that in human–robot interaction some people suffer from tension emotion, which would disturb expression of other autonomic emotions, such as joy. For example, a person under tension would not make a big smile. This will interfere with the ability of robots to detect subtle nuances of utterance. In summary, we hypothesize that if there is strong self-reported emotion expressed, it is difficult to recognize autonomic emotion.

7.2.2.3 Hypothesis Verification

We conducted a preliminary experiment for a verification of the hypothesis. Forty-five university students participated as subjects. In the experiment, participants talked with Robovie 50 cm distant from it. It repeatedly asks some simple questions, such as "Let's play together, shall we?" "Do you think Robovie is cute?" Subjects were required to answer the questions (1) freely, (2) positively, and (3) negatively. We recorded subjects' faces observed from its camera for later analysis.

7.2.2.3.1 Labeling of Facial Expression

We selected 72 data items from obtained faces where the subjects reported they expressed their emotions. Third persons other than authors and participants rated the recorded faces with 7 scales related to emotions: anger, disgust, fear, joy, sadness, surprise, and tension. These seven emotions were chosen because Ekman's six basic emotions are recognizable by the facial emotion recognition system we used for Robovie and the effect of tension emotion is what we want to verify. The rating was conducted with −3 (not match at all) to 3 (match very much) scales for each emotion (f_{anger}, $f_{disgust}$, f_{fear}, f_{joy}, f_{sad}, $f_{surprise}$, $f_{tension}$), and the most highly rated emotion was selected for the label of the face. (For example, if f_{joy} was highest, the face was labeled "joy.")

7.2.2.3.2 Results for Hypothesis Verification

We classified all rated items into two tension-related categories: *with tension* ($f_{tension} > 0$) and *without tension* ($f_{tension} \leq 0$), and two categories about the attitude of messages: *positive* and *negative* (Table 7.1). As a result, 23 items are classified in the *with tension* category among the 72 items. That is, tension emotion is observed in 32% of the items. Most cases in the *with tension* category were labeled as tension emotion because it was highly compared with other emotions. It reminds us of the importance of moderating tension emotions in human–robot interaction.

We believe that this result verifies our hypothesis. Often, people become tense and express tension emotion, which is a kind of self-reported emotion. It suppresses other autonomic emotions. Thus, *with tension* cases, it is difficult to guess the nuances of a message from observed emotions.

On the contrary, in *without tension* cases, we often observed joy emotion in *positive* answer cases, and anger and disgust emotions in *negative* answer

TABLE 7.1

Result of Preliminary Experiment

Emotion	$f_{tension} > 0$ (with Tension)		$f_{tension} \leq 0$ (w/o Tension)	
	Positive	*Negative*	*Positive*	*Negative*
No. of items	14	9	34	15
Anger	0%	0%	6%	27%
Disgust	3%	0%	22%	45%
Fear	0%	7%	0%	0%
Joy	29%	7%	60%	27%
Sad	6%	0%	12%	1%
Surprise	0%	0%	0%	0%
Tension	62%	86%	0%	0%

cases. It suggests that we can observe the nuances of messages from emotions. For example, if the result from speech recognition on a message is ambiguous between "yes" and "no," positive emotions lead us to believe that the message is related to a positive answer.

7.2.3 System Configuration

Our system has two features. First, there is tension–moderation interaction (shown in Section 7.2.3.6). The system senses tension emotion, and when this occurs, the robot behaves to moderate the tension. Second, if tension moderation successfully moderates users' tension, users will not be disturbed by tension emotions and express other automatic emotions. Thus, the system detects other emotions to supplement insufficient speech recognition ability (shown in Section 7.2.3.5).

7.2.3.1 Overview

Figure 7.2 gives an overview of the developed system. It consists of three recognition units: face tracking unit, speech recognition unit, and emotion recognition unit. The face tracking unit keeps tracking the face of an interacting person so that it can observe facial emotions and direct its own directional microphone to him or her. The emotion recognition unit detects tension emotion and other emotions, which is used for the behavior selection and speech recognition units, respectively. If no tension emotion is detected, the result from the speech recognition unit is used for behavior selection.

7.2.3.2 Robovie

A humanoid robot "Robovie" is used. It is capable of humanlike expression and recognizes individuals by using various actuators and sensors. Its body

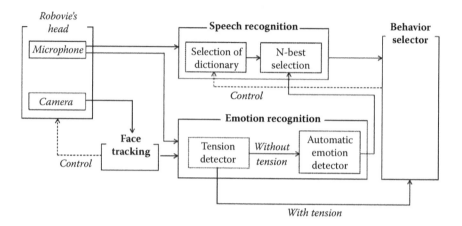

FIGURE 7.2
Robovie's speech-based interaction system.

possesses highly articulated arms, eyes, and a head, which were designed to produce sufficient gestures to communicate effectively with humans. The sensory equipment includes auditory, tactile, ultrasonic, and vision sensors, which allow the robot to behave autonomously and to interact with humans. All processing and control systems, such as the computer and motor control hardware, are stored inside the robot's body.

7.2.3.3 Face Tracking Unit

We used a technique presented in [11] to track a face, which integrates information from the eye-camera and omnidirectional camera. It controls the direction of Robovie's head so that it can keep watching his or her face. Through this process, it obtains his or her frontal face, which is used in the emotion recognition unit. At the same time, since it controls Robovie's head toward him or her, the attached directional microphone turns toward him or her, which results in less noisy auditory input from him or her.

7.2.3.4 Emotion Recognition Unit

There are two sources for the emotion recognition unit: facial emotions and vocal emotions. Facial emotions are recognized by using a system developed by Littlewort et al. [12]. It is based on Ekman's FACS (Facial Action Coding Systems) and outputs likelihoods of each of six emotions (anger, disgust, fear, joy, sadness, and surprise) with SVM (support vector machine) so that we can recognize these six emotions and the neutral emotion based on these likelihoods.

The vocal emotions are recognized based on the method of Komatani et al. [6]. It uses 29 features, which are calculated based on a fundamental frequency (F0), power, length of utterance, and duration between utterances. It detects joy and perplexity emotion with SVM. Further, we added detection for tension emotions with C5.0 decision tree with same 29 characteristics. Trained with 400 data items obtained in an experiment with 15 subjects with the same settings as those reported in Section 7.2.2, it shows performance for tension emotion detection as 67.1% of correct answers for test data.

7.2.3.5 Speech Recognition Unit

7.2.3.5.1 Situated Recognition on Speech Recognition

We adopted a speech recognition system, Julian [13], which allows us to switch its dictation dictionaries and grammars. Based on our constructive approach with situated recognition [14], a dictionary and grammar is chosen along with the current situation Robovie generated. For example, when it asks for the name of a place, such as "where are you from?" it uses a dictionary that includes the names of places so that it can get better recognition results in a noisy environment. Each dictionary contains 50 to 200 words.

It outputs *N-best* results ($1 \leq N \leq 5$) of recognition with a certain threshold on likelihood score. These *N-best* results are compared with the output from emotion recognition.

7.2.3.5.2 Supplementing Speech Recognition from Emotion Recognition

We supplemented the deficient ability of the robot's speech recognition with the results from emotion recognition. This supplement is conducted when tension emotion is not detected and it expects positive–negative answer, such as an answer for yes–no question.

The result from emotion detection is classified into three categories: positive emotion (denoted as *Pe*), negative emotion (*Ne*), and neutral emotion (*Nt*). If there is a joy emotion from either facial or vocal emotion recognition, the system classified it as positive emotion (*Pe*). If there is anger, disgust, fear, or sadness from facial expression, or perplexity from vocal information detected, the system classified it as negative emotion (*Ne*). Otherwise, the system assumes it as neutral emotion (*Nt*). If conflict occurs between the recognition in facial and vocal emotions, it uses the one detected as facial emotion.

We decided this classification by referring to the analysis results reported in Section 7.2.2. In addition, although there were no cases of fear and sadness related to negative utterances, we classified fear and sadness emotions as negative emotion, because there was not enough data to conclude that these emotions are not related to negative utterances, while these emotions are usually related to negative situation.

As a result, if positive (negative) emotion is detected, it refers to the *N-best* results from speech recognition, and chooses the words with positive (negative) meanings. It means that it chooses the word that fits better with nuances estimated from nonverbal information. The meaning of words, whether positive or negative, is defined in advance.

TABLE 7.2

Robot's Utterances for Normal Behaviors and Tension-Moderating Behaviors Used in the Experiment

Normal Behaviors		Tension-Moderating Behaviors	
N1	Hello	T1	I'm Robovie. What is your name?
N2	Let's talk together. Shall we?	T2	I'm from ATR. Where are you from?
N3	Let's play together. Shall we?	T3	What do you think today's weather?
N4	Do you think Robovie is your friend?	T4	Let's play a game of paper-scissor-rock. Shall we? (It plays the game.)
N5	Do you think Robovie is cute?	T5	Do you know the song of "A Flower Smile"? (It sings the song.)
N6	Bye-bye		

7.2.3.6 Behavior Selector: Tension Moderation

Robovie exhibits interactive behavior, such as shaking hands, greeting persons, and asking some simple questions [15]. We extended the selection of behavior to moderating tension emotion.

The mechanism of tension moderation is quite simple. There is a *behavior selector* that selects "tension-moderating" behavior if tension emotion is recognized in the emotion recognition unit. Otherwise, the *behavior selector* chooses interactive behaviors, and used integrated recognition results from the speech recognition unit and emotion recognition units to switch to the next interactive behaviors.

So far, we have implemented a simple example of tension-moderating behavior. Table 7.2 shows the behaviors used in this study. It includes self-introduction and talking about weather, which humans often do with a person they meet for the first time. Otherwise, it chooses its normal interactive behaviors, and uses the result from speech recognition, which is supplemented with the result from emotion recognition.

7.2.4 Experiment

We conducted an experiment to verify the effect of the developed system.

7.2.4.1 Settings

[Participants] The participants in our experiment were 27 university students (12 men and 15 women). Their average age was 19.7 years.

[Methods] The experiment was conducted in a room in our laboratory, in which the participants and Robovie talked. Each participant stood about 50 cm from Robovie. At first, Robovie showed normal behavior "hello" (N1, shown in Table 7.2). If there was a tension emotion detected in the participant's response to "hello," it started a tension-moderating behavior (T1); otherwise, it initiated the next normal behavior (N2). After it spoke a sentence as listed in Table 7.2, Robovie expected the response from the participant, and it spoke shortly in reply to the response, such as "thank you," "I'm glad," and "it's disappointing," according to the result from speech recognition. After that, if it detected a tension emotion from the response, it performed the next tension-moderating behavior; otherwise, it performed the next normal behavior. For example, after it performed S2 after N2, it executed S3 or N3, which was decided by referring the results for tension detection. The experiment lasted after the execution of the last normal behavior (N6). If the last tension-moderating behavior (T5) was performed, even if it detected a tension emotion, it performed the next normal behavior.

[Measurement] We videotaped the scene of the experiment to record the faces and utterances of participants. The results of recognition in Robovie were also recorded for later analysis. After the experiment, we asked the participants to answer the following yes-no questions:

Q1. "Did Robovie understand your utterances?"

Q2. "Do you feel it is easier to speak to Robovie after this session, compared to before the session?"

Q3. "Are there problems regarding communication with Robovie?"

7.2.4.2 Results

First of all, we ask a third person to label emotions of participants during each of their utterances in the experiment. There were three classes (positive emotion: *Pe*, negative emotion: *Ne*, neutral: *Nt*) for facial emotions and vocal emotions, and two classes for tension (*with tension, without tension*). As a result, there were 220 utterances of 27 participants analyzed. It is used as a ground truth of emotion recognition to evaluate performance of the developed system.

7.2.4.2.1 Results for Performance of Emotion Recognition

We compared the emotion recognition output from the system with the labeled emotions. Table 7.3 shows the results of the comparison, where "success rate" represents the rate that the output from the system correctly matched the labeled emotions for all classes among 220 utterances. As a result, the success rate for the tension detection (denoted as *Tension* in Table 7.3), emotion detection from face (*Facial*), and that of from vocal (*Vocal*) was 55.0%, 36.8%, and 31.7%, respectively.

Further, we analyzed the detailed failure rate. In Table 7.3, "opposite," "false neutral," and "false from neutral" represent the rate of each case among all 220 utterances. As a result, the labeled results and the output from the system were often mismatched around the boundary on neutral emotion, which lowered the success rate of the system. On the other hand, the number of items in the "opposite" case is relatively small (11.8% for facial and 6.5% for vocal).

TABLE 7.3

Result for Emotion Recognition

		Tension	Facial	Vocal
No. of classes		**2**	**3**	**3**
No. of analyzed data items		170	202	170
No. of error data items (omitted from analysis)		50	8	50
Success rate		550%	368%	317%
Failed rate	**Opposite:**			
	Pe (Ne), classified as Ne (Pe)	—	11.8%	6.5%
	False neutral:			
	Pe, Ne, classified as Nt	—	19.8%	6.5%
	False from neutral:			
	Nt, classified as Pe, Ne	—	31.6%	55.3%

Meanwhile, eight data items for facial emotion caused errors and were omitted from this analysis. Since participants sometime looked away or their faces were sometime occluded by their arm, Robovie could not observe the participants' faces and caused error output for facial emotion recognition. Also, 50 data items for vocal emotion were omitted due to errors in the low-level analysis program for retrieving F0 and pitch, which is probably due to background noise in inputs.

7.2.4.2.2 Moderation of Tension

Next, we compared the effects of tension-moderating behaviors for moderating tension emotion with that of normal behaviors (these behaviors used for the experiment are shown in Table 7.2). Table 7.4 shows the result of the comparison. "Success rate of tension-moderation" represents the rate of the disappearance of tension emotion after the execution of a behavior when tension emotion was observed before the execution. For example, 50 cases of tension emotion were observed before the execution of tension-moderating behavior, while 21 cases of tension emotion were observed after these behaviors. These evaluations are based on the labeled emotions.

We also compared the effects of tension-moderating behaviors for improving positive–negative emotions with that of normal behaviors. Table 7.4 also shows the result of the comparison regarding the improvement of emotions, where "success rate of improving emotion-expression" represents the rate of

TABLE 7.4

Effect of Tension-Moderating Behaviors

	Normal Behavior	Tension-Moderating Behavior
Success rate of tension-moderation	12% (7/54)	42% (21/50)
Success rate of improving emotion-expression	31% (15/48)	54% (19/35)

TABLE 7.5

Results for Supplementing Speech Recognition from Emotion Recognition

	No. of Utterances	Supplementation With	Supplementation W/O
Success rate All utterances for answering positive (yes)–negative (no) questions	136	70%	60%
Only the utterances where participants expressed positive or negative emotions	68	50%	43%

the appearance of positive or negative emotions after the execution of a behavior when no positive or negative emotions were observed before the execution.

7.2.4.2.3 Improvement of Speech Recognition

Table 7.5 shows the results of the effect of supplementing speech recognition from emotion recognition. There were 136 utterances in reply to the Robovie's "yes"–"no" questions for participants (N2, N3, N4, N5, T4, and T5 in Table 7.2). We analyzed the performance of speech recognition for these utterances, because the supplementation mechanism now only works for utterances where interacting person answers either positively or negatively.

In total, Robovie detected the correct answer for 70% utterances with the supplementation mechanism, but for 60% utterances without it. We defined the correct answer as detection of the correct keyword in the utterance, such as "yes," "ok," "cute," "let's play," among 8–9 sets of keywords. Furthermore, we focused on utterances where the participants expressed positive or negative emotions. As shown in the table, the performance for these utterances was 50% (34/68) with supplementation, while performance without supplementation was 43% (29/68). There are three cases where the supplementation mechanism for speech recognition failed.

7.2.4.2.4 Subjective Evaluation

Participants answered the questionnaire after the experiment. As a result, 20 of the 27 participants answered that Robovie understood their utterances (Q1: understandings), and 24 participants answered that communication with Robovie became easier as they communicated with it (Q2: easiness), as shown in Table 7.6. Chi-square test proved that the number of participants who answered "yes" for Q1 and Q2 are statistically more than that of the

TABLE 7.6

Results for Subjective Evaluation

	Yes	No	
Q1. Understandings	20	7	$p < .05$
Q2. Easiness	24	3	$p < .01$
Q3. Difficulty	11	16	n.s.

participants who answered "no," which seems to suggest that majority of participants enjoyed the communication with Robovie.

On the other hand, 11 participants had problems in communicating with Robovie. Their comments were "it was difficult to communicate, once I recognized it as a machine," "it was difficult to expect that it would talk," "I don't think it understands what I say," etc. This seems to suggest that its communication abilities are still far from those of humans.

7.2.4.3 Discussion

As we intended, tension-moderating behavior has an effect for moderating the tension emotion of interacting person. Moreover, it has the effect of improving their expression of positive or negative emotions, which fits the model proposed in Section 7.2.2.4 (Figure 7.2). The supplementation mechanism also worked well to improve the performance of the speech recognition. Questionnaire results also showed that most participants enjoyed the communication with Robovie.

On the other hand, the success rates for emotion recognition were relatively poor, and far lower than their original performances. Since the failure rate for the "opposite" case was not so large, we believe one major difficulty was detecting subtle expressions, while noisy input also probably decreased performance. Often, an emotion recognition system is trained and evaluated with very expressive examples in a noise-free environment, such as a face with a big smile and a voice in an anechoic room. However, our practical use for the robot highlighted the weakness of the emotion recognition system for subtle expression in a noisy environment.

We believe that the lower failure rate of emotion recognition in the "opposite" case seems to also explain why the supplementation mechanism worked with poor success rate in emotion recognition. When speech recognition works well, N-best result from speech recognition only includes few candidates with higher likelihood score, and so the poor result from emotion recognition does not affect it much. On the contrary, when the output from speech recognition is ambiguous, both positive and negative words are included in the N-best candidates. Since emotion recognition does not often fall in the "opposite" case, it rather improves the performance of speech recognition.

One interesting finding was that speech recognition performance was low for the utterances where participants expressed their emotions (43%, without the supplementation mechanism, in Table 7.5). Because the speech recognition system is usually trained for the utterances with neutral emotions, perhaps it does not work well in such situations without any special training.

Regarding the tension-moderating mechanism, it worked well for this particular experiment, because almost all participants were under tension at the beginning. Thus, although the tension detection unfortunately behaved nearly randomly, Robovie sometimes exhibited tension-moderating behaviors, and, as a result, moderated their tension. We believe that the effects of tension-moderating behaviors suggest the usefulness of our framework; however, for practical and effective use, we should improve the performance of tension detection so that it does not unnecessarily exhibit tension-moderating behaviors.

7.2.5 Conclusion

This chapter reports one of our early works aimed at adapting a robot's behavior to a user's context. As our analysis revealed, users sometimes exhibit tension emotion, which prevents them from exhibiting other emotions. The developed system exhibits tension-moderating behavior when tension emotion is observed; otherwise, it uses observed emotions to supplement speech recognition results. The experimental results revealed that the proposed mechanism works reasonably well.

Among the studies presented in the book, perhaps this specific study is less rigorous with regard to experimental evaluation, as it was conducted in the early stage in our series of the project. Nevertheless, we included this study in the book because the study highlighted the important aspect of adaptation in human–robot interaction, that is, the awareness of users' negative emotion. Our field trials started to reveal people who show positive attitude toward robots. However, there are some people who hesitate to interact. To make social robots really socially acceptable, we believe that the adaptation capability toward people who hesitate to interact with, or perhaps to accept, robots should be developed. We believe this study is one of the pioneering works for this adaptation.

Acknowledgments

We wish to thank Prof. Masuzou Yanagida at Doushisha University and Prof. Tatsuya Kawahara at Kyoto University for their valuable advices. This research was supported by the National Institute of Information and Communications Technology of Japan.

References

1. Mehrabian, A. (1980), *Silent Messages*, Thomson Learning College.
2. Takanishi, A., Sato, K., Segawa, K., Takanobu, H., and Miwa, H. (2000), An anthropomorphic head-eye robot expressing emotions based on equations of emotion, *IEEE International Conference on Robotics and Automation (ICRA2000)*, pp. 2243–2249.

3. Breazeal, C., and Scassellati, B. (1999), A context-dependent attention system for a social robot, *International Joint Conf. on Artificial Intelligence (IJCAI1999)*, pp. 1146–1153.

4. Breazeal, C., and Aryananda, L. (2002), Recognizing affective intent in robot directed speech, *Autonomous Robots*, 12:1, pp. 83–104,.

5. Baek, S.-M., Tachibana, D., Arai, F., and Fukuda, T. (2004), Situation based task selection mechanism for interactive robot system, *IEEE/RSJ International Conference on Intelligent Robots and Systems (IROS 2004)*, pp. 3738–3743.

6. Komatani, K., Ito, R., Kawahara, T., and Okuno, H. G. (2004), Recognition of emotional states in spoken dialogue with a robot, *Int. Conf. on Industrial and Engineering Applications of Artificial Intelligence and Expert Systems (IEA/AIE'04)*, pp. 413–423.

7. Ekman, P. (1992), An argument for basic emotions, *Cognition and Emotions*, pp. 169–200.

8. Russell, J. A. (1980), A circumplex model of affect, *Journal of Personality and Social Psychology*, pp. 1161–1178.

9. Oatley, K., and Jenkins, J. M. (1996), *Understanding Emotions*, Oxford, UK, Blackwell.

10. Nomura, T., Kanda, T., and Suzuki, T. (2004), Experimental investigation into influence of negative attitudes toward robots on human-robot interaction, *SID (Social Intelligence Design)*.

11. Shiomi, M., Kanda, T., Miralles, N., Miyashita, T., Fasel, I., Movellan, J., and Ishiguro, H. (2004), Face-to-face interactive humanoid robot, *IEEE/RSJ International Conference on Intelligent Robots and Systems (IROS2004)*, pp. 1340–1346.

12. Littlewort, G., Bartlett, M. S., Fasel, I., Chenu, J., Kanda, T., Ishiguro, H., and Movellan, J. R. (2003), Towards social robots: Automatical evaluation of human-robot interaction by face detection and expression classification, *International Conference on Advances in Neural Information Processing Systems*, 1438–1441.

13. Lee, A., Kawahara, T., Takeda, K., Mimura, M., Yamada, A., Ito, A., Itou, K., and Shikano, K. (2002), Continuous speech recognition consortium—an open repository for CSR tools and models—, *IEEE Int'l Conf. on Language Resources and Evaluation*.

14. Kanda, T., Ishiguro, H., Imai, M., and Ono, T. (2004), Development and evaluation of interactive humanoid robots, *Proceedings of the IEEE*, vol. 92, No. 11, pp. 1839–1850.

7.3 Adapting Nonverbal Behavior Parameters to Be Preferred by Individuals

Noriaki Mitsunaga,[1] Christian Smith,[2] Takayuki Kanda,[1]
Hiroshi Ishiguro,[1,3] and Norihiro Hagita[1]

[1]*ATR Intelligent Robotics and Communication Laboratories* [2]*Royal Institute of Technology, Stockholm* [3]*Graduate School of Engineering, Osaka University*

ABSTRACT

A human subconsciously adapts his or her behaviors to a communication partner in order to make interactions run smoothly. In human–robot interactions, not only the human but also the robot is expected to adapt to its partner. Thus, to facilitate human–robot interaction, a robot should be able to read subconscious comfort and discomfort signals from humans and adjust its behavior accordingly, just as a human would. However, most previous research works expected the human to consciously give feedback, which might interfere with the aim of interaction. We propose an adaptation mechanism based on reinforcement learning that reads subconscious body signals from a human partner, and uses this information to adjust interaction distances, gaze meeting, and motion speed and timing in human–robot interaction. We use gazing at the robot's face and human movement distance as subconscious body signals that indicate a human's comfort and discomfort. An evaluation trial with a humanoid robot which has ten interaction behaviors has been conducted. The experimental result of 12 subjects shows that the proposed mechanism enables autonomous adaptation to individual preferences. Also, a detailed discussion and conclusions are presented.

7.3.1 Introduction

How does a social robot become familiar with individual users? In Chapter 2, we show a case in which a robot serves a user repeatedly for long-term interaction in a shopping mall. There should be other working contexts in which a robot would interact with individuals repeatedly. For example, in a hospital, a robot would be employed to serve patients [1]. In a supermarket, a robot would be used to assist elderly people's daily shopping [2]. Apparently at home a robot would meet family member every day. It is obvious that if a robot does everything in the same way as before, it will be perceived as "robotlike."

A previous study has established the importance of adapting robots' verbal behavior in long-term interaction. There are previous studies on methods to change [3] and increase variety [4] of verbal utterance as well as the method to increase exhibited familiarity [5] over time; however, it was not known how to adapt non-verbal behavior of a robot.

In contrast, in human communication studies, it is evident that people adapt their nonverbal behavior to their partners. For instance, people keep shorter distances when they are familiar [6]. Different people tend to meet the gaze of others to different extents [7]. People's movements are aligned if they intend to be friendly [8].

Moreover, we can see subtle signals in human communications that make it possible for people to adjust their nonverbal behavior. For example, when a conversational partner stands too close, we tend to move away, and when we are stared at, we tend to avert our eyes [9].

This chapter reports a study that aims to make a social robot adapt its non-verbal behavior to an individual partner. There are a number of adaptation studies in human–robot interaction, though such studies require direct input from a human user. The difficulty in adaptation in human–robot interaction is to observe users' feedback. For this problem, we propose a behavior adaptation system based on *policy gradient reinforcement learning* (PGRL). Using comfort and discomfort signals from the human partner as inputs for the reward function, the system simultaneously searches for the behavioral parameters that maximize the reward, thereby also, respectively, maximizing and minimizing the actual comfort and discomfort experienced by the human. We use a reward function that consists of the human's movement distance and gazing period in human–robot communication [10]. We use six behavioral parameters: three parameters that determine interaction distance/personal space [6], and one parameter each to determine the time that the robot looks at the human's face, the delay after which the robot makes a gesture after an utterance, and the speed of the gestures.

7.3.2 The Behavior Adaptation System

7.3.2.1 *The Robot and Its Interaction Behaviors*

We used an interactive humanoid robot, Robovie. It is placed in a room equipped with a 3D motion capture system comprising 12 cameras. The system captures 3D positions of markers attached to the human and the robot at a sampling rate of 60 Hz. We used a space measuring 3.5 × 4.5 m in the middle of the room. The data from the system were forwarded to the robot via Ethernet, resulting in data lags of at most 0.1 s, ensuring sufficient response speed.

As a simple example, there are 10 interactive behaviors used in the study, which were also used in our previous study [11], listed as follows:

a. Ask to hug it (hug)
b. Ask to touch it (ask for touch)
c. Ask for a handshake (handshake)
d. Rock-paper-scissors
e. Exercise
f. Pointing game
g. Say "Thank you. Bye-bye. See you again" (monologue)
h. Ask where the human comes from (ask where from)
i. Ask if it is cute (ask if cute)
j. Just look at the human (just looking)

Figure 7.3 shows a scene in which these behaviors were used in an interaction. In the figure, I, P, and S indicate intimate, personal, and social distance classes

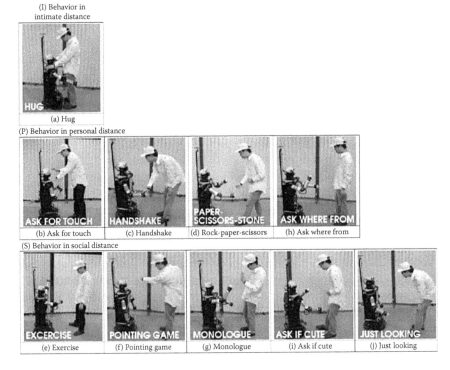

FIGURE 7.3
Ten interactive behaviors used in the study.

of the behaviors, respectively. We classified interaction behaviors into Hall's three interaction categories [6]: *intimate* (0 to 0.45 m), *personal* (0.45 to 1.2 m), and *social* (1.2 to 3.6 m) by a preliminary trial. (We exposed eight subjects to the behaviors, and let them choose what distance they were comfortable with for each of these. The robot did not move its wheels in the prestudy.) Each behavior took about 10 s to run. The robot always initiates the interaction and like a child asks to play, since it has been designed as a childlike robot.

7.3.2.2 Adapted Parameters

We adopted six parameters to be adapted by the system as follows:

Interaction distances for three classes of proxemics zones: *intimate*, *personal*, and *social* distances. During the interaction, the robot tries to keep the interaction *distance* of the category to which the behavior belongs. The *distance* was measured as the horizontal distance between the robot and human foreheads.

Gaze-meeting ratio: The extent to which the robot would meet a human's gaze. The robot meets the human's gaze in a cyclic manner, where the robot meets the human's gaze and breaks eye contact in cycles that last 0 to 10 s (randomly determined, average 5 s), as this is the average cycle length for gaze meeting and averting in human–human interaction [7]. The parameter *gaze-meeting ratio* is the portion of each cycle spent meeting the human participants's gaze.

Waiting time : Waiting time between utterance and gesture. The *waiting time* controlled how long the robot would wait between utterance and action. When it performs behaviors from (a) *hug* to (g) *monologue* that require motion on the part of the human, the robot starts actions after it makes an utterance (like "Please hug me," "Let's play rock-paper-scissors," and so on) and *waiting time* has passed.

Motion speed: The *motion speed* controlled the speed of the gesture and motion. If *motion speed* is 1.0, the gesture/motion is carried out at the same speed that the gesture/motion is designed for.

As for *gaze-meeting ratio, waiting time*, and *motion speed*, the same values are used for all interaction behaviors. We chose these since they seem to have a strong impact on interaction and low implementation costs, allowing us to keep the number of parameters small and thereby the dimensionality of the search space.

7.3.2.3 Reward Function

An analysis of human body movement [10] in human–robot interaction revealed that people's subjective evaluation had a positive correlation with the length of the *gazing* time and a negative correlation with the distance that the participants moved. We built a reward function based on this previous study. To be specific, it is based on the *movement* distance of the human and the proportion of time spent *gazing* directly at the robot in one interaction.

Figure 7.4 shows a block diagram of reward calculation. The foreheads' positions and directions of the human and the robot are measured by a 3D motion capture system at a sampling frequency of 60 Hz. They are then projected onto the horizontal plane and down-sampled to 5 Hz. The human's *movement* distance is the sum of the distances that he or she moved in all sampling periods (200 ms) of a behavior. The *gazing* factor was calculated as the percentage of time that the participant's face was turned toward the robot

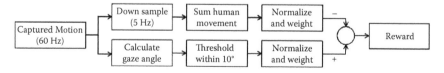

FIGURE 7.4
Computation of reward function.

in the interaction behavior, with an allowance of ±10 degrees in horizontal direction. The reward function R is defined as

$$R = -0.2 \times (movement\ distance\ [\text{mm}])$$

$$+ 500 \times \frac{(\text{time human spent looking at robot})}{(\text{time spent for the interaction behavior})}$$

The weights in the reward function were determined with a preliminary trial. We recorded the human and the robot movements, and measured the parameters the human preferred. We then ran the PGRL algorithm with different weights using the recorded values and tuned weights so that the behavioral parameters quickly and stably converged to the preferred values. Note that this process did not require much tuning.

7.3.2.4 The PGRL Algorithm

From the reinforcement-learning algorithms, we choose the PGRL method, which is known to be quickly convergent, because the learning space could be considerably reduced by using human knowledge when we prepare the policy function.

Figure 7.5 shows the algorithm we adopted [12]. The Θ indicates the current policy or the values of n behavioral parameters. A total of T perturbations of Θ are generated, tested with a person, and the reward function is evaluated. Perturbation Θ^t of Θ is generated by randomly adding Ej, 0, or −Ej to each element θj in Θ. The step sizes Ej are set independently for each parameter.

The robot tests each policy Θ^t with an interaction behavior and then receives the reward. Note that the interaction behavior can be different for each test since we assume that the reward is not dependent on the behaviors but on the policy only. When all T perturbations have been run, the gradient A of the reward function in the parameter space is approximated by calculating the partial derivatives for each parameter. Thus, for each parameter θj, the average reward when Ej is added, no change is required, and cases when Ej is subtracted are calculated. The gradient in dimension j is then regarded as 0 if the reward is greatest for the unperturbed parameter, and is considered to be the difference between the average rewards for the perturbed parameters otherwise. When the gradient A has been calculated, it is normalized to overall step size η and for the individual step sizes E in each dimension. The parameter set Θ is then adjusted by adding A.

7.3.3 Evaluation

We evaluated the adaptation system with human participants. They interacted with the robot for 30 min; during the interaction, the robot adapted behavioral parameters based on the proposed algorithm. We analyzed the degree to which the parameters were adjusted to their preferred parameters.

```
1   Θ ← Initial parameter set vector of size n
2   ε ← parameter step size vector of size n
3   η ← overall step size
4   while (not done)
5       for t = 1 to T
6           for j = 1 to n
7               r ← unbiased random choice
                    from {−1, 0, 1}
8               θ_j^t ← θ_j + ε_j * r , where Θ^t is
                perturbed parameter set of same size as Θ
9           for t = 1 to T
10              Run system using parameter set Θ^t,
                evaluate rewards
11          for j = 1 to n
12              Avg_{+ε,j} ← average reward for all Θ^t
                    with positive perturbation in dimension j
13              Avg_{0,j} ← average reward for all Θ^t
                    with zero perturbation in dimension j
14              Avg_{−ε,j} ← average reward for all Θ^t
                    with negative perturbation in dimension j
15              if (Avg_{0,j} > Avg_{+ε,j}) AND
                    (Avg_{0,j} > Avg_{−ε,j})
16                  a_j ← 0
17              else
18                  a_j ← ( Avg_{+ε,j} − Avg_{−ε,j} )
19          A ← A/|A| * η
20          a_j ← a_j * ε_j, ∀ j
21          Θ ← Θ + A
```

FIGURE 7.5

This is the PGRL algorithm that we adopted for the adaptation system.

7.3.3.1 Participants

There are 15 participants (9 males and 6 females) whose ages were between 20 and 35. All participants were employees or interns of our lab. However, they were not familiar with the experimental setup, and most had no prior experience of the type of interaction used in the study. None of them had taken part in the prestudy.

7.3.3.2 Settings

The participants were asked to stand in front of the robot and to interact with it in a relaxed, natural way. The robot randomly selected one of the ten interaction behaviors. After one behavior finished, it randomly selected the next one. The interaction session lasted for 30 min. For simplicity, except for controlling the selection not to repeat the same behavior twice in a row, we did not pay any special attention to the randomness of the selection.

During the interaction, the adaptation system was running on the robot in real time. Table 7.7 shows the initial values and the search step sizes. The initial values were set slightly higher than the parameters that the participants in preliminary trials preferred.

TABLE 7.7

Initial Values and Step Sizes of Behavioral
Parameters

#	Parameter	Initial Value	Step Size
1	Intimate distance	0.50 m[a]	0.15 m
2	Personal distance	0.80 m	0.15 m
3	Social distance	1.0 m	0.15 m
4	Gaze-meeting ratio	0.7		0.1
5	Waiting time	0.17 s	0.3 s
6	Motion speed	1.0	0.1

[a] We used 0.50 m distance although it is outside the
intimate range.

For the duration of each interaction behavior, or the test of a policy, the robot kept the interaction distance and other parameters according to Θ^t. The reward function was calculated for each executed interaction of the robot using the accumulated motion and gaze-meeting percentage for the duration of the behavior. The duration was measured from just after the robot selected the behavior or before it speaks, and it ends at the end of the behavior or just before the next behavior selection. A total of ten different parameter combinations were tried before the gradient was calculated and the parameter values updated ($T = 10$). The participants did not notice the update of the policy during interaction since the calculation was done instantaneously.

7.3.3.3 Measurements

After the 30 min interaction session, we measured the following behavioral and subjective measures.

Preference regarding interaction distances: The participants were asked to stand in front of the robot, at the distance he or she felt was most comfortable for a representative action for each of the three distances studied by using behaviors (a) *hug*, (c) *handshake*, and (g) *monologue*, respectively. Other parameters—*gaze-meeting-ratio, waiting time,* and *motion speed*—were fixed to 0.75, 0.3, and 1.0 s, respectively. We also asked the participants to indicate acceptable limits, that is, how close the robot could come without the interaction becoming uncomfortable or awkward, as well as how faraway the robot could be without disrupting the interaction.

Preference regarding gazing, waiting, **and** *speed:* Each participant was shown the robot's behavior performed in turn with three different values—low, moderate, and high—for each of the parameters *gaze-meeting-ratio, waiting time,* and *motion speed.* The parameters that were not measured at this time were fixed to *moderate* values (same for all subjects). The subjects were asked to indicate which of the three displayed behaviors they felt comfortable with. A few subjects indicated several values for a single parameter, and some indicated preferences between or outside the shown values. We recorded as such if he or she said so.

The *moderate* value for gazing, 0.75, was based on the average preference in the preliminary trial; the *high* value was set to continuous gazing at 1.0; and the *low* value was set equidistantly from the moderate value at 0.5. For motion speed, the preprogrammed motions were assumed to be at a moderate speed, and the *high* and *low* values were set to be noticeably faster and slower, respectively, than this value. A time of 0 s was chosen as the *low* value for waiting, and the *moderate* value of 0.5 s was chosen to be a noticeable pause, and the *high* value of 1 s was chosen to be noticeably longer.

We used (g) *monologue* to measure *gaze-meeting-ratio* and *motion speed*. The interaction distance was fixed at 1.0 m. We used (a) *hug* to measure *waiting time* and asked the participants to stand at a comfortable distance since the *intimate* parameter varied among the subjects. Other parameters were fixed at the averaged values that the subjects in the prestudy preferred.

7.3.4 Adaptation Results

For most participants, at least some of the parameters reached reasonable convergence to stated preferences within 15–20 min, or approximately 10 iterations of the PGRL algorithm. We have excluded the results of three participants who neither averted their gaze nor shifted their position however inappropriate the robot's behavior became, but showed their discomfort in words and facial expression to the experimenter. These trials had to be aborted early as safe interaction could not be guaranteed, so no usable data could be collected from them.

Figure 7.6 shows the learned values for the distances as compared to the stated preferences for 12 participants, excluding the 3 participants above. The intimate distance converged in the acceptable range for 8 out of 12 participants. The personal and social distances converged in acceptable range for 7 and 10 participants, respectively. The learned distance is here calculated as the average parameter value during the last quarter (about 7.5 min) of each run, since the algorithm keeps searching for the optimum value. The bars show the interval for acceptable distance and the preferred value, and the asterisks "*" are the learned values.

Figure 7.7 shows the remaining three parameters, where circles "○" indicate what values the participants indicated were preferred. Some participants indicated a preference between two values, and these cases are denoted with a triangle "∇" showing that preferred value. The asterisks again show the learned values as the mean values for the last quarter of the trials. The *gaze-meeting-ratio, waiting time*, and *motion speed* converged to the values near selected values for 7, 6, and 5 out of 12 participants, respectively.

As can be seen, most of the parameters converged in the acceptable ranges; however, there is a large difference in the success rate between different parameters. This is because not all parameters are equally important for

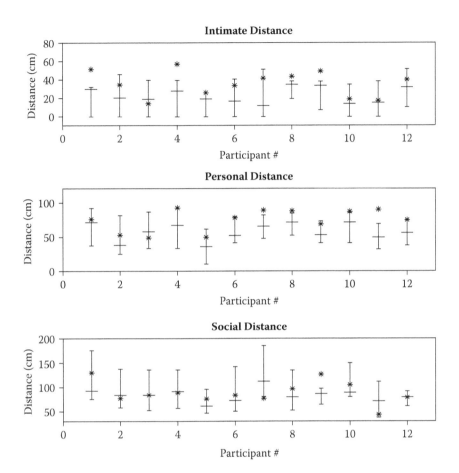

FIGURE 7.6
Learned distance parameters and preferences for 12 participants.

successful interaction. It is a typical trait for PGRL that parameters having a greater impact on the reward function are adjusted faster, while those with a lesser impact will be adjusted at a slower rate.

Figure 7.8 shows the average deviation for each parameter over all participants at the initial quarter and during the last quarter of the experiments. All values have been normalized for step size Ej. Most parameters converged to within one step size, the exceptions being the *personal* and *social* distance parameters. It should be noted that for these parameters the average stated tolerance (the difference between the closest comfortable distance and the farthest) was of a size corresponding to several step sizes. For example, for *personal* distances, the average stated tolerance was 3.0 step sizes and for *social* distances it was 5.0. As Figure 7.6 shows, for all participants except one, the learned *social* distance parameter values fall within the stated acceptable interval. Further details are presented in [13].

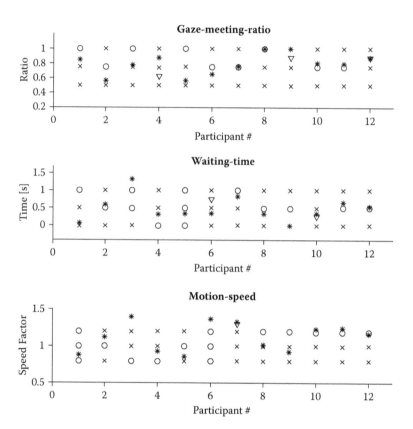

FIGURE 7.7
Earned parameters for *gaze-meeting-ratio*, *waiting-time*, and *motion-speed*.

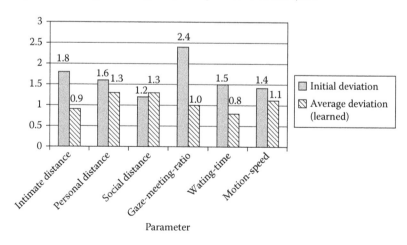

FIGURE 7.8
Average deviations from preferred values (normalized to step size units).

7.3.5 Discussion

7.3.5.1 Difficulties

We consider this study topic challenging, thus it is useful to note the difficulties we faced. First, it is very difficult to measure true preferences. For example, participant 5 was satisfied with the learned parameters even though they were far from the stated preferences. On the contrary, participant 7 claimed that the robot got too close, even though the distances were greater than her stated preferences.

Second, for some participants, the method could not find the gradient of some parameters nor the direction to the local optimum. The reason is that the behaviors of the participants did not display any difference for policies Θ^t if the deviation of the current parameter was too great from the preferred values. This suggests that the adaptation should be done in an appropriate search space, where participants behave as the reward function expects. This would further enforce the design goal "to not behave unacceptably during adaptation."

Third, there were participants whose behaviors differed from our expectations. Participant 3 had a tendency to fix her gaze on the robot when the *motion speed* was higher than her preference. Thus, we need different reward functions for people who have different reactions.

We have simplified the human model by making the following two assumptions. First, people will not stand still nor stare at the robot to show discomfort or confusion since the interactions are simple to understand and the adapted parameters start from points where they do not produce anti-social behavior. Second, we have only chosen actions that contain a direct interaction and communication between the robot and the participants in our trials. How to prepare an appropriate reward function based on gaze information when these assumptions are not fulfilled remains an issue for the future. There also are many other possible human behaviors that can be used for rewarding, as [14] suggests.

7.3.6 Conclusion

We have proposed a behavior adaptation system based on PGRL for a robot to interact with a human. We have shown that the robot has successfully adapted at least part of the learning parameters to individual preferences for 11 out of the 12 participants in the trial. Although there are a number of limitations for generalization such as the fact that adaptation was only successful in the limited context of interaction, we consider that this study revealed a possibility for a robot to adapt its nonverbal behavior to individual users. It is particularly notable that the robot adapted its behavior through the observation of people's subtle behavioral response. We believe this is an important step toward making robots socially acceptable in daily repeated interactions.

Acknowledgments

This research was supported by the Ministry of Internal Affairs and Communications of Japan.

References

1. Pineau, J., Montemerlo, M., Pollack, M., Roy, N., and Thrun, S. (2003), Towards robotic assistants in nursing homes: Challenges and results, *Robotics and Autonomous Systems*, vol. 42, pp. 271–281.
2. Iwamura, Y., Shiomi, M., Kanda, T., Ishiguro, H., and Hagita, N. (2011), Do elderly people prefer a conversational humanoid as a shopping assistant partner in supermarkets?, *6th ACM/IEEE International Conference on Human-Robot Interaction (HRI2011)*, pp. 449–456.
3. Gockley, R., Bruce, A., Forlizzi, J., Michalowski, M., Mundell, A., Rosenthal, S., Sellner, B., Simmons, R., Snipes, K., Schultz, A. C., and Wang, J. (2005), Designing robots for long-term social interaction, *IEEE/RSJ Int. Conf. on Intelligent Robots and Systems (IROS2005)*, pp. 1338–1343.
4. Kanda, T., Sato, R., Saiwaki, N., and Ishiguro, H. (2007), A two-month field trial in an elementary school for long-term human-robot interaction, *IEEE Transactions on Robotics*, vol. 23, pp. 962–971.
5. Kanda, T., Shiomi, M., Miyashita, Z., Ishiguro, H., and Hagita, N. (2010), A communication robot in a shopping mall, *IEEE Transactions on Robotics*, vol. 26, pp. 897–913.
6. Hall, E. T. (1966), *The Hidden Dimension*. Doubleday Publishing.
7. Duncan, S. Jr., and Fiske, D. W. (1977), *Face-to-Face Interaction: Research, Methods, and Theory*. Lawrence Erlbaum Associates.
8. Chartrand, T. L., and Bargh, J. A. (1999), The chameleon effect: The perception-behavior link and social interaction, *Journal of Personality and Social Psychology*, vol. 76, pp. 893–910.
9. Sundstrom, E., and Altman, I. (1976), Interpersonal relationships and personal space: Research review and theoretical model. *Human Ecology*, vol. 4, no. 1, pp. 47–67.
10. Kanda, T., Ishiguro, H., Imai, M., and Ono, T. (2003), Body movement analysis of human-robot interaction. In *Int. Joint Conference on Artificial Intelligence (IJCAI 2003)*, pp. 177–182.
11. Kanda, T., Ishiguro, H., Imai, M., and Ono, T. (2004), Development and evaluation of interactive humanoid robots, *Proceedings of the IEEE*, vol. 92, pp. 1839–1850.
12. Kohl, N., and Stone, P. (2004), Policy gradient reinforcement learning for fast quadrupedal locomotion. In *Proceedings of International Conference on Robotics and Automation*, vol. 3, pp. 2619–2624, IEEE.
13. Mitsunaga, N., Smith, C., Kanda, T., Ishiguro, H., and Hagita, N. (2008), Adapting robot behavior for human-robot interaction, *IEEE Transactions on Robotics*, vol. 24, pp. 911–916.
14. Tickle-Degnen, L., and Rosenthal, R. (1990), The nature of rapport and its nonverbal correlates. *Psychological Inquiry*, vol. 1, no. 4, pp. 285–293.

7.4 Learning Pedestrians' Behavior in a Shopping Mall

Takayuki Kanda, Dylan F. Glas, Masahiro Shiomi,
Hiroshi Ishiguro, and Norihiro Hagita

ATR Intelligent Robotics and Communication Laboratories

ABSTRACT

This chapter introduces our technique to learn pedestrians' behavior from their trajectories. A large number of people walk in public spaces daily without giving the task much conscious thought. We humans have a common-sense understanding about our environment. For example, we know how spaces are used, and how people typically walk in these spaces. We can tell who would be interested in shopping. We present a series of techniques for anticipating people's behavior in a public space, mainly based on the analysis of accumulated trajectories, and we demonstrate the use of these techniques in a social robot. In a field test, we demonstrate that this system enables the robot to serve people efficiently.

7.4.1 Introduction*

When we plan to use a robot in a public environment such as a shopping mall, what is the preparation required? Some information should be loaded into the robot. For instance, a robot needs a map to recognize its location, and to know the important places, such as the locations of shops. Perhaps such map information might be already available, as maps are typically prepared for human visitors.

Moreover, it is very important for a robot to understand information about people's behavior. In particular, since timing is highly critical for social interactions, anticipating the motion and behavior of customers will be a fundamental capability for a robot. For example, if a robot is designed to invite customers to a shop, it should approach people who are walking slowly and possibly window shopping. To approach those customers, we first anticipate areas where a behavior (e.g., walking slowly) is often observed. We then use a technique based on global behavior estimation to "preapproach" customers who are most likely to exhibit the chosen behavior. This chapter reports a

---

* This chapter is a version of a previously published paper (Kanda, T., Glas, D. F., Shiomi, M., Ishiguro, H. and Hagita, N., 2008, Who will be the customer? A social robot that anticipates people's behavior from their trajectories, Int. Conf. on Ubiquitous Computing (UbiComp2008), pp. 380–389, and Kanda, T., Glas, D. F., Shiomi, M., and Hagita, N., 2009, Abstracting People's Trajectories for Social Robots to Proactively Approach Customers, *IEEE Transactions on Robotics*, vol. 25, pp. 1382–1396), modified to fit the context of this book.

study to make it possible for a robot to anticipate people's behavior, which is based on the learning of people's daily behavior in the environment.

7.4.1.1 Related Works

This study considers three essential types of trajectory-related information: local behavior, the use of space, and global behavior. We define the term *local behavior* to refer to basic human motion primitives, such as walking, running, going straight, and so on. The observation of these local behaviors can then reveal information about the *use of space*, that is, how people's behavior differs in different areas of the environment. Finally, for more insight into the structure of people's behaviors, we look at *global behavior*, that is, overall trajectory patterns comprising several local behaviors in sequence, such as "entering through the north door, walking across the room, and sitting at the desk." Global behaviors are highly dependent on environments.

The detection of local behaviors and analysis of the use of space can be valuable in anticipating where behaviors are statistically likely to occur; however, an analysis of global behavior is far more powerful for predicting *individual* behavior. As people using the space have a variety of goals, an understanding of global behavior is essential in enabling the robot to anticipate the future behaviors of individuals.

7.4.1.2 The Use of Space

Information on the general use of space has also been retrieved. Nurmi et al. applied a spectral clustering method for identifying meaningful places [1]. Aipperspach et al. applied clustering to UWB sensor data to identify typical places in the home [2]. Koile et al. conducted a clustering of spaces with a focus on the relationships between velocity and positions, which enabled a partitioning of space into "activity zones." For example, places for walking, working, and resting were separated [3]. Our work involves partitioning space in a similar manner, but based on position and local behavior. In addition, we also consider how the distribution of these zones varies as a function of time.

7.4.1.3 Global Behavior

People's overall behavior has been studied in some degree, such as people's goals and intentions [4]. In a museum context, Sparacino developed the "museum wearable," where people were classified into three visiting patterns. Depending on the pattern, the system adjusted the way it presented information [5]. This is a good example of the use of global behavior; however, the places and the model of global behaviors were carefully prepared by a human designer.

In contrast, we have applied a clustering technique to identify typical visiting patterns in a museum without providing any environmental information [6]. One of the novel points of our work is that the designer of the system provides information only about the *target local behavior*, with no knowledge about the structure of the space or of people's global behaviors. In addition to the previous work, this paper provides a method of online estimation of global behavior, which is indispensable for providing services.

The online estimation of global behaviors is difficult as, by definition, any global behavior being observed in real time is unfinished and thus not completely observable. Thus, it is necessary to estimate the true global behavior from a limited dataset. Krumm et al. developed a technique they call "Predestination," which enables someone's driving destination to be estimated [7]. Liao et al. developed a technique for a person wearing GPS to infer her destination, transportation mode, and anomalous behavior [8].

While personal history of previous destinations was an important part of those studies, our anticipation technique for the shopping arcade assumes zero knowledge of a given person's individual history. Our technique is predicated on our observations of tens of thousands of people and the expectation that a new person's global behavior will be similar to those previously observed.

The concept of behavior anticipation is not without precedent in robotics. For example, Hoffman et al. demonstrated the value of anticipatory action in human–robot collaboration [9]. However, our use of global behaviors is a unique approach to behavior anticipation in this field.

7.4.1.4 System Configuration

Figure 7.9 shows the overview of the system. The details will be explained in the following sections.

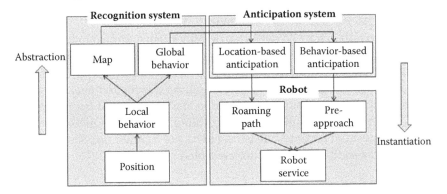

FIGURE 7.9
Overview of the system.

Laser range finder

FIGURE 7.10
The shopping arcade and laser range finders.

7.4.1.4.1 Position

We conducted our experiments in a popular entertainment and shopping arcade located by the entrance to Universal Studios Japan, a major theme park. We operated the robot within a 20 m section of the arcade, with shops selling clothing and accessories on one side and an open balcony on the other. The motion of people through this area was monitored using a ubiquitous sensor network consisting of six SICK LMS-200 laser range finders mounted around the perimeter of the trial area at a height of 85 cm (Figure 7.10).

A particle filtering technique was used to track people's trajectories through this space. The location of each person in the scan area was calculated based on the combined torso-level scan data from the laser range finders.

In our tracking algorithm, a background model is first computed for each sensor, by analyzing hundreds of scan frames to filter out noise and moving objects. Points detected in front of this background scan are grouped into segments, and segments within a certain size range persisting over several scans are registered as human detections.

Each person is then tracked with a particle filter, using a linear motion model with random perturbations. Likelihood is evaluated based on the potential occupancy of each particle's position (i.e., humans cannot occupy spaces that have been observed to be empty). By computing a weighted average across all of the particles, x-y position is calculated at a frequency of approximately 37 Hz. This tracking technique provides quite stable and reliable position data, with a position accuracy measured to be +/− 6 cm for our environment. Further details on this algorithm are presented in [10].

7.4.1.4.2 Local Behavior

As defined earlier, "local behaviors" represent basic human motion primitives. We began our analysis with the classification system, which uses

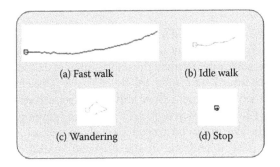

FIGURE 7.11
Example trajectories for local behaviors.

SVM (support vector machine) to categorize trajectories based on their velocity, direction, and shape features.

To include a wide variety of movement types, we initially defined a set of 20 local behavior classes, based on walking style and walking speed, considering both 5-s trajectory segments and short-duration (2 s) segments. Each class has about 200 samples for learning, consisting of 2- or 5-s trajectory segments. We manually labeled these samples. The classification method averaged 89.6% accuracy for category estimation. The following are examples of the categories used:

Style: Classes describing style of motion: *straight, right-turn, left-turn, wandering, U-turn,* and *stop.*

Speed: Classes describing overall speed: *run, fast-walk, idle-walk, stop,* and *wait.*

In the subsequent analysis, we merged several local behavior classes for simplicity. Within "Style," the classes *left-turn, right-turn,* and *U-turn* were all merged into the *wandering* category. Within "Speed," we merged *stop* and *wait* into the *stop* category. We also merged classes for short duration and 5 s behavior. Thus, we reduced the set to the following four local behaviors: *fast-walk, idle-walk, wandering,* and *stop.* Figure 7.11 shows examples of these local behaviors. We define the position P_t^n of visitor n at time t to include the x-y coordinates (x, y) as well as Boolean variables indicating the presence or absence of local behavioral primitives $P_{fast-walk}$, $P_{idle-walk}$, $P_{wandering}$, P_{stop}.

7.4.1.5 Analysis of Accumulated Trajectories

Based on the position and local behavior data thus obtained, an analysis was performed to obtain a higher-level understanding of the use of space and people's global behaviors. This analysis constitutes the foundation for the robot's ability to anticipate people's local behaviors.

7.4.1.5.1 Data Collection

Human motion data was collected for a week in the shopping-arcade environment, from 11 a.m.–7 p.m. each day, including 5 weekdays and 2 weekend days. We chose this time schedule because the shops open at 11 a.m., and the number of visitors drops after 7 p.m., after the theme park closes in the evening.

In this environment, the major flow consisted of customers crossing the space from the left to the upper right or vice versa, generally taking about 20 s to go through. We removed trajectories shorter than 20 s, in order to avoid noise from false detections in the position tracking system. In all, we gathered 11,063 visitor trajectories.[*]

7.4.1.5.2 Use of Space (Map)

The first analysis task was to identify how the space was used, and how the use of space changed over time. We applied the ISODATA clustering method to achieve this. First, we partitioned the time into 1-h segments categorized as weekday or weekend. We then partitioned the space into a 25-cm grid, with each grid element containing histogram data of local behaviors: $H_{fast-walk}(i,t)$, $H_{idle-walk}(i,t)$, $H_{wandering}(i,t)$, and $H_{stop}(i,t)$, where $H_x(i,t)$ denotes the number of occurrences of local behavior x at time slice t within grid element i, which is normalized for each local behavior x.

To make the data set more manageable, we first combined time slices based on their similarity. The difference between time slices t_1 and t_2 is defined as

$$\sum_i \sum_x |H_x(i,t_1) - H_x(i,t_2)|$$

We then combined spatial grid cells where the distance was smallest and the grid was spatially connected. The distance between grid cells i and j is defined as

$$\sum_t \sum_x |H_x(i,t) - H_x(j,t)|$$

As is usual for this type of explorative clustering, we arbitrarily set the number of partitions to intuitively understand the phenomena occurring in the environment. We set the number of spatial partitions to be 40 and temporal partitions to be 4. Figure 7.12 shows a visualized output of the analysis. The partitions are color-coded according to the dominant local behavioral primitive in each area. Blue (medium gray on monochrome printouts) represents the areas where the *fast-walk* behavior occurred more frequently than

---

[*] In this study, we obtained approval from shopping mall administrators for this recording under the condition that the information collected would be carefully managed and only used for research purposes. The experimental protocol was reviewed and approved by our institutional review board.

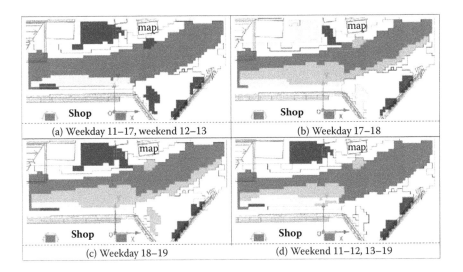

FIGURE 7.12
Analysis of the use of space.

any other local behaviors. Thus, people tend to pass directly through this area, which can be thought of as "corridor" space.

The areas where the *idle-walk* primitive occurred most frequently are colored with green (or light gray).

In some areas, the use of space was very clearly observed to change as a function of time. The lower-left area is in front of a shop. When the shopping arcade was busy in the evening, as in Figure 7.12b, with people coming back from the theme park, many people were observed to slow down in front of the shop, and the "corridor" space changed into "in front of shop" space with *idle-walk* becoming dominant; however, when there were not so many people, such as midday during the week as in Figure 7.12a, these areas disappeared and became similar to other "corridor" space. The lower-right side of the map represents the side of the corridor, where people tend to walk slowly when the arcade is busy (Figure 7.12b,c).

The areas where the *stop* primitive was most frequent are colored with dark brown (or dark gray). In Figure 7.12, these areas can mainly be found in the upper center (photo: Figure 7.13a) and the bottom right (photo: Figure 7.13b). These areas contain benches, and can be considered "rest space."

In the upper center area, below the word *map*, there is a small space where *stop* is the dominant primitive in Figure 7.12a, whereas *idle-walk* is dominant in Figure 7.12b through d. A map of the shopping arcade is placed on that wall. Customers sometimes slowed down, stopped, and looked at this map (Figure 7.13c). The statistical analysis clearly revealed this phenomenon as defining a distinct behavioral space.

The areas where the *wandering* primitive was dominant are colored pink (or very light gray). All maps in Figure 7.12 show the space immediately

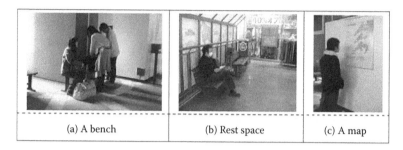

| (a) A bench | (b) Rest space | (c) A map |

FIGURE 7.13
Examples of the actual use of the space.

in front of the shop as having this property. The areas where none of the primitives were dominant, such as the bottom-right space, are colored white. These areas were not used so much.

To summarize, we have demonstrated that through this analysis technique, we can separate space into semantically meaningful areas such as the corridor, the space in front of the shop, the area in front of the map, and the rest space. It also reveals how usage patterns change over time, such as the change of dynamics in the space in front of the shop.

7.4.1.5.3 Global Behavior

To identify typical global behaviors of shopping arcade visitors, we classified trajectories with a k-means method [6]. For the clustering, the distance between two trajectories is computed by using a DP matching method. Figure 7.14 shows how the comparison of trajectories works.

For the DP matching, we again partitioned the space into a 25-cm grid, to easily compare trajectories. As a result, the environment was separated into 2360 grid elements. The DP matching method was chosen for its simplicity and the fact that it does not require particular tuning of parameters. Since global behaviors naturally emerge through the interactions between people and their environment, we believe that it is best to minimize the number of parameters that need to be adjusted manually, keeping the process simple and generalizable.

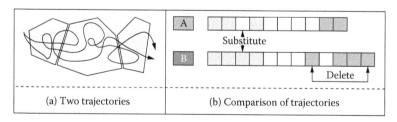

| (a) Two trajectories | (b) Comparison of trajectories |

FIGURE 7.14
Comparison of trajectories based on DP matching.

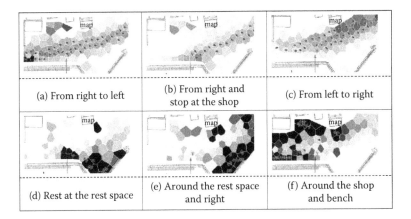

(a) From right to left	(b) From right and stop at the shop	(c) From left to right
(d) Rest at the rest space	(e) Around the rest space and right	(f) Around the shop and bench

FIGURE 7.15
Six typical patterns of global behavior.

The trajectories are segmented into 500 ms time steps, and they are compared with each other based on the physical distance between them at each time step. To this is added a cost function, based on "insert" and "delete" operation costs in the DP matching, where we defined the cost of a single insertion or deletion to be 1.0 m.

Figure 7.15 shows a visualization of the global behaviors at $k = 6$. We separated the space into 50 similarly sized partitions by the k-means method for this visualization, although the actual computation used 2360 partitions. In the figure, each area is colored according to its dominant local behavior primitive. For example, blue represents *fast-walk*, and green represents *idle-walk*. Solid colors indicate a frequency of occurrence of at least one standard deviation above average, and lighter tints represent weaker dominance, down to white if the frequency is more than a standard deviation below average. Frequent transitions between adjacent areas are shown by arrows.

The following six typical global behaviors were retrieved:

(a) Pass through from right to left (7768 people)

This pattern represents one of the major flows of people, who are coming back from the theme park (on the right) on their way to the train station (on the left). In this pattern, most of the areas are colored blue because the most frequent primitive in those areas was *fast-walk*. In front of the shop, there are some areas colored green, which represent spaces where people slow down to look at the shop.

(b) Come from the right, and stop at the shop (6104 people)

In this pattern, people come from the right side and enter the shop, as these trajectories mostly disappear at the shop.

(c) Pass through from left to right (7123 people)

This is also a major pattern, where people are coming from the train station and going in the direction of the theme park. In contrast to the patterns in (a) and (b), people rarely stopped or slowed down in front of the shop.

(d) Rest at the rest space (213 people)

In this pattern, people mostly spent time in the bottom-right rest space (Figure 7.13b) where benches were placed.

(e) Around the rest space and right (275 people)

Similar to the pattern in (d), but people moved around the right area more, and not around the shop area.

(f) Around the shop and bench (334 people)

People mainly came from the left side, walking slowly, and stopped in front of the shop as well as in front of the map.

In summary, this analysis technique has enabled us to extract typical global behavior patterns. These results show that most people simply pass through this space, while a smaller number of people stop around the rest space or the map area. People tend to stop at the shop more often when they come from the right, a result which makes intuitive sense, as the shopping arcade is designed mainly to attract people coming back from the theme park.

7.4.1.6 Anticipation System

Robots differ from other ubiquitous computing systems in that they are mobile, and it takes some time for a robot to reach a person in need of its service. Thus, the ability to anticipate people's actions is important, as it enables the robot to pre-position itself so it can provide service in a timely manner.

We assume here that the robot's service is targeted toward people who are performing some particular local behavior, such as *stop* or *idle-walk*. The robot system uses the results of the analysis about the use of space and global behavioral primitives to anticipate the occurrence of this "target behavior." At the same time, the robot system tries to avoid people who are performing particular local behaviors, such as *fast-walk*, which we refer to as "nontarget behavior." To anticipate local behaviors, we use two mechanisms: location-based anticipation and behavior-based anticipation.

7.4.1.6.1 Location-Based Anticipation

As shown in Figure 7.12, the system has use-of-space information about the frequency of the local behaviors associated with spatial and temporal partitions. The robot uses this information to estimate the locations in which people will be statistically likely to perform the target behavior.

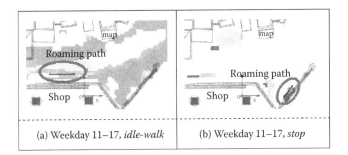

FIGURE 7.16
Example of anticipation map.

Figure 7.16 shows an example anticipation map. The red areas represent areas where the system anticipates both a high likelihood of the target behavior and a low likelihood of the nontarget behavior. The robot roams through this high-likelihood area looking for people. The black line on the map represents its automatically generated roaming path.

In one scenario, the robot's task might be to invite people to visit a particular shop. In this case, selecting *idle-walk* as the target behavior and *fast-walk* as the nontarget behavior might be appropriate, since the robot wants to attract people who have time and would be likely to visit the store. Figure 7.16a is the anticipation map for this scenario, calculated for the behavior patterns observed on weekdays, between 11 a.m. and 5 p.m. Several areas away from the center of the corridor are colored, and the roaming path is set in front of the shop.

In a different scenario, the robot's task might be to entertain idle visitors who are taking a break or waiting for friends. Particularly because this shopping arcade was situated near a theme park, this is quite a reasonable expectation. In this case, it would be more appropriate to select *stop* as the target behavior and *fast-walk* as the nontarget behavior. Figure 7.16b is the anticipation map for this second scenario. In this case, only a few areas are colored. The roaming path is set to the bottom-right area.

Note that the roaming path was automatically calculated based on the anticipation map. No additional knowledge about the space was provided by designers.

7.4.1.6.2 Behavior-Based Anticipation

The second technique used for anticipating local behaviors is to estimate the global behaviors of people currently being observed, and then to use that information to predict their expected local behaviors a few seconds in the future.

To ensure prediction accuracy, we used a large number of clusters for the global behavior analysis. We clustered the human motion data collected earlier into 300 global behavior patterns. Next, to predict the global behavior of a new trajectory that has been observed for T seconds, the system compares the new trajectory with the first T seconds of the center trajectory of

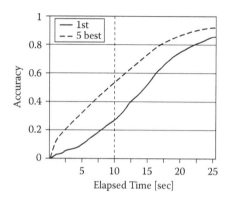

FIGURE 7.17
Accuracy of the prediction of global behavior.

each of the 300 clusters, using the same DP matching technique applied earlier for deriving the global behaviors. The cluster with the minimum distance from the new trajectory is considered to be the best-fit global behavior for that trajectory.

Figure 7.17 shows the prediction accuracy for observed trajectories from 0 to 25 s in length. Here, we used 6 of the 7 days of data to create the prediction model, and tested its ability to predict the remaining 1 day of the accumulated data. The accuracy accounts for only trajectories of total length greater than 20 s, as we filtered out shorter trajectories for calculating global behaviors. The result labeled "1st" represents the case where the best-fit global behavior at time T was the correct one (the cluster the trajectory finally fit with at completion). The result labeled "5 best" is the result if we define success to mean that correct global behavior falls within the top 5 results. Performance levels off after 20 s. Since there are 300 global behaviors, we believe that a success rate after 10 s of 45% and after 15 s of 71% for "5 best" represents fairly good performance.

After the most likely global behaviors are selected, the person's future position and local behavior are predicted based on an "expectation map." An expectation map is a data structure prepared a priori for each global behavior. For each 500 ms time step along the trajectories, a 25-cm grid representation of the observed space is added to the map. Each element of this grid contains likelihood values for each of the four local behaviors to occur in that location at anytime *after* that time step. These likelihood values are empirically derived from the original observed trajectories falling within the chosen global behavior cluster, and they represent the average frequency of the occurrence of each local behavior after that time step. We used the 5-best result to create the expectation map for the person by accumulating each expectation map of 5-best global behaviors.

Figure 7.18 shows expectation maps for various time increments. The solid circles represent the positions of people walking through the space, with

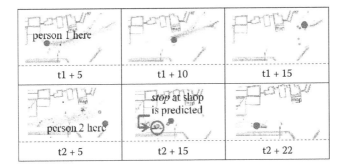

FIGURE 7.18
Example of prediction of future behaviors.

the person of interest outlined in red. The expectation map for that person's estimated global behavior is shown, where the area colored blue represents the area where *fast-walk* is expected, and the green area represents the area where *idle-walk* is expected. The three figures in the top row show the trajectory for person 1, who was first observed at time t_1. The first figure shows time $t_1 + 5$ s, where the expected local behaviors can be seen tracing a path through the corridor, heading toward the upper right. In fact, this course was correctly predicted, and the person followed that general path. The second line is the trajectory for person 2, first observed at time t_2. Here, since the person walked slowly, it predicted the course to the left with *idle-walk* behavior. At time t_2+15, it started to predict the possibility of *stop* at the shop, which finally came to be true at time t_2+22.

We measured the accuracy of position prediction for four time windows: 0–5, 5–10, 10–15, and 15–20 s in the future. Predictions were begun after a trajectory had been observed for 10 s, as the estimation of global behavior is not stable until then. We again used 6 days of data from the accumulated trajectories to predict the data of the remaining day. Our method predicts the future position as the center of mass of the expectation map. Figure 7.19 compares our method with position prediction based on the velocity over the last second. As the velocity method cannot account for motions such as following the shape of the corridor, our method performs about twice as accurately.

We then measured the correctness of the system's predictions of the future positions and local behaviors for each person, evaluated in four places (indicated by 3 m circles in Figure 7.20) where qualitatively distinct behaviors were observed in the use-of-space analysis. For each place, at each moment, the system predicted whether the person would exhibit each of the local behaviors at that place for forecast windows of 0–5, 5–10, 10–15, and 15–20 s.

Figures 7.21 and 7.22 show the system's prediction performance. In each figure, the left graph shows the accuracy of the prediction for the case where the target local behavior occurred at each place, and the right graph shows the accuracy of the prediction where the behavior did not occur. We define the occurrence of the local behavior as the case where the person appeared

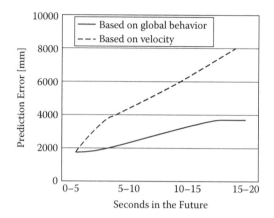

FIGURE 7.19
Prediction accuracy for position.

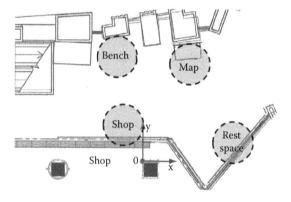

FIGURE 7.20
Places used for measuring the performance.

at the place in the predicted 5-s window (e.g., between 5 s and 10 s), and performed the target local behavior more than other local behaviors. The accuracy value used for each person is the average across all predictions made for that person, and the value shown in the graph is the average across all people.

Figure 7.21 shows that the prediction was fairly accurate for the *stop* behavior, particularly at the bench and the rest space. Prediction was 92% accurate at the bench even for 15–20 s in the future, while nonoccurrence was predicted with 88% accuracy. This good performance was because people who stay in these areas often stay for a long time. Results were more marginal at the map and shop, with 62% accuracy for occurrence and 63% for nonoccurrence predicted at the shop for 0–5 s in the future. For 15–20 s in the future, the performance is still marginal, with 48% accuracy for occurrence and 71% for nonoccurrence predicted at the shop.

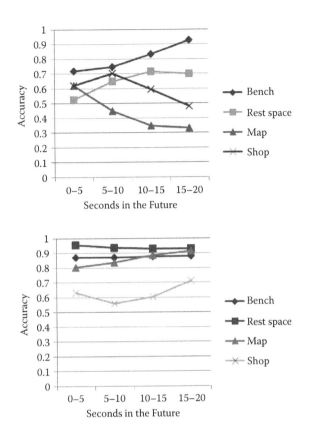

FIGURE 7.21
Prediction accuracy for *stop* behavior. (Above: behavior occurred at the place, Bottom: behavior did not occur at the place.)

In contrast, as Figure 7.22 shows, the system predicted *idle-walk* with high accuracy 0–5 s ahead at the map and the shop. Even for 15–20 s ahead, the system was able to predict 33% of the occurrences at the shop as well as 86% of the nonoccurrences, which we consider to be a good result, as it is rather difficult to predict walking behavior in the future. The prediction of occurrence was not successful at the rest space, as the system mostly predicted nonoccurrence, since *idle-walk* rarely happened there.

Regarding the remaining two behaviors, for *wandering* the system predicted over 50% of occurrences and 85% of nonoccurrences for 0–5 s ahead at all four places. For the 15–20 s window, it predicted 73% of occurrences and 93% of nonoccurrences at the bench but not so well for the map and shop. It predicted *fast-walk* at the map and shop well until 10 s; for example, it predicted 86% of occurrences and 60% of nonoccurrences at the shop for 5–10 s in the future, though it does not predict the future well beyond 10 s.

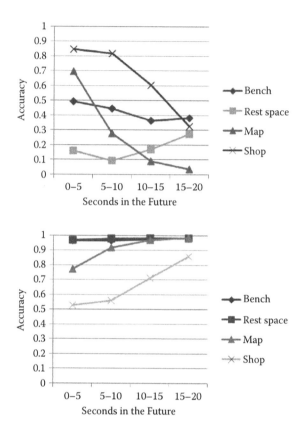

FIGURE 7.22
Prediction accuracy for *idle-walk* behavior. (Above: behavior occurred at the place, Bottom: behavior did not occur at the place.)

We believe these anticipation results are useful for the robot. The robot is designed to wait for people in areas where it anticipates frequent occurrence of the target behavior. Behavior-based anticipation performs particularly well in areas where the anticipated behaviors occur often, such as *stop* near the benches and rest space, and *idle-walk* in the corridor in front of the map and shop. As these are the areas predicted by the location-based anticipation method, the two anticipation techniques complement each other nicely.

7.4.1.7 Field Test with a Social Robot

In this section, we show examples where a social robot provides services using our system. A human designer defines the contents of the service as well as the context in which the robot should provide the service. Here, the notable point is that the designer only specifies the target local behavior, such as "stopping." The robot system then automatically computes the information about space and global behavior so that the robot can efficiently wait for people in promising areas, and then proactively approach people who are anticipated to exhibit the target local behavior.

For these services a robot has an advantage over mobile devices, in that people do not need to carry any hardware; however, there is the additional challenge that robots need to approach the person quickly enough to start the service. For this purpose, anticipation plays an important role.

7.4.1.7.1 *Robot Hardware*

We used a humanoid robot Robovie, characterized by its humanlike physical expressions. It is 120 cm high and 40 cm in diameter. It is equipped with basic computation resources, and it communicates with the ubiquitous sensor network via wireless LAN.

7.4.1.7.2 *An Example of Application*

We tested the technique with a simple application. The robot is used to entertain visitors in the form of chatting. As mentioned earlier, the shopping arcade is next to an amusement park, so it is a reasonable for the robot to be entertaining people who have free time. Other potential applications would be seen in other chapters in the book (e.g., in Chapter 2, we reported the use of social robots as museum guides and advertisement providers, and also for direction giving.)

The chat was about the attractions in the amusement park. For example, the robot says, "Hi, I'm Robovie. Yesterday, I saw the Terminator at Universal Studios. What a strong robot! I want to be cool like the Terminator. 'I'll be back... .'" We set the target local behavior as *stop*, and nontarget as *fast-walk*, in order to serve people who are idle.

We conducted a field trial to investigate the effectiveness of the system. Based on the anticipation mechanism and its current position, the robot set its roaming path near the bench and waited for a person to approach. When the robot predicted that a detected person would probably exhibit the *stop* behavior, the robot began positioning itself near her general area (preapproach). When she came in front of the shop, she stopped (partly, we assume, because she was intending to stop regardless of the robot, and partly because she noticed the robot approaching her). Once she stopped, the robot approached her directly, and they had a chat. This is a typical pattern illustrating how people and the robot started to interact. Overall, people seemed to enjoy seeing a robot that approached them and spoke.

To evaluate the performance, we compared the situation with the developed system "with anticipation," and "without anticipation," and measured how much the anticipation mechanism improved the efficiency. In the without-anticipation condition, the robot simply approached the nearest person engaged in *stop* behavior. We measured the performance for one hour in total for each condition. We prepared several time slots and counterbalanced the order.

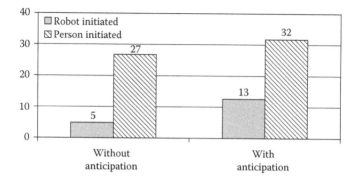

FIGURE 7.23
The number of services provided.

Figure 7.23 shows the number of people to whom the robot provided services. Due to the novelty of the robot, people often initiated interactions on their own; in such cases, the anticipation mechanism is irrelevant. Thus, we classified the robot's interactions into two categories. The first case, "robot-initiated," is the situation where the robot initiated the service by approaching the person and entering into conversation distance. Thus, the number of robot-initiated services indicates how the robot's anticipation system improved the efficiency of the service. The second case, "person-initiated," is the situation where the person approached the robot while it was talking to someone else. While the robot was talking with two girls, a child came from the left. When the girls left, the child stood in front of the robot to start talking with it.

The results in Figure 7.23 indicate that the number of robot-initiated services "with anticipation" is much larger than "without anticipation." In other words, anticipating enables the robot to provide the service more efficiently. Due to the novelty factor of the robot, the number of person-initiated services is quite large. We believe that in the future, when robots are no longer so novel to people, there will be fewer person-initiated interaction, and the results concerning anticipation will become much more significant.

7.4.2 Conclusion

This chapter reports a study to learn information about people's walking behavior from their trajectories. From a large number of recorded trajectories, it analyzes how people use the space, and how people overall go through, that is, global behavior in the environment. Such information is used by a social robot to anticipate people's future behavior, so that a robot can effectively position itself to serve a potential user enough before his or her arrival. The demonstration in this study might be only a small application; nevertheless, we consider that it shows how useful the information about people's walking behavior is. A capacity of anticipation is revealed to be important in other studies as well [11].

Acknowledgments

We wish to thank Dr. Miyashita, Mr. Nishio, and Mr. Nohara for their help. This research was supported by the Special Coordination Funds for Promoting Science and Technology of the Ministry of Education, Culture, Sports, Science and Technology, the Japanese Government.

References

1. Nurmi, P., and Koolwaaij, J. (2006), Identifying meaningful locations, In *Proc. Mobiquitous 2006*, 1–8.
2. Aipperspach, R., Rattenbury, T., Woodruff, A., and Canny, J. (2006), A quantitative method for revealing and comparing places in the home, In *Proc. Ubicomp 2006*, 1–18.
3. Koile, K. et al. (2003), Activity zones for context-aware computing, In *Proc. Ubicomp 2003*, 90–106.
4. Chai, X., and Yang, Q. (2005), Multiple-goal recognition from low-level signals, In *Proc. AAAI-05*, 3–8.
5. Sparacino, F. (2002), The museum wearable, In *Proc. Museums and the Web (MW2002)*.
6. Kanda, T. et al. (2007), Analysis of people trajectories with ubiquitous sensors in a science museum, *In Proc. Int. Conf. on Robotics and Automation (ICRA2007)*, 4846–4853.
7. Krumm J., and Horvitz, E. (2006), Predestination: Inferring destinations from partial trajectories, In *Proc. Ubicomp2006*, 243–260.
8. Liao, L., Patterson, D., Fox, D., and Kautz, H. (2007), Learning and inferring transportation routines, *Artificial Intelligence*, 171(5–6), 311–331.
9. Hoffman, G., and Breazeal, C. (2007), Effects of anticipatory action on human-robot teamwork, *In Proc. Int. Conf. on Human-Robot Interaction (HRI2007)*, 1–8.
10. Glas, D. et al. (2007), Laser tracking of human body motion using adaptive shape modeling, *In Proc. Int. Conf. Intelligent Robots and Systems (IROS2007)*, 602–608.
11. Satake, S., Kanda, T., Glas, D. F., Imai, M., Ishiguro, H., and Hagita, N. (2009), How to approach humans?: Strategies for social robots to initiate interaction, *ACM/IEEE Int. Conf. on Human-Robot Interaction (HRI2009)*, pp. 109–116.

Index